THE ATOMS OF LANGUAGE

THE
ATOMS
OF
LANGUAGE

MARK C. BAKER

Basic Books
A Member of the Perseus Books Group

Published by Basic Books,
A Member of the Perseus Books Group.

Design by Rachel Hegarty. Set in 11 point Sabon.

Library of Congress Cataloging-in-Publication Data
Baker, Mark C.
    The atoms of language / Mark C. Baker.—1st ed.
        p.   cm.
    Includes bibliographical references and index.
    ISBN 0-465-00521-7
    1. Language and languages.   I. Title.
P107 .B35   2001
400—dc21   2001035936

FIRST EDITION

01 02 03 04 / 10 9 8 7 6 5 4 3 2 1

For Linda,
Who does so much more than bake,
and who is so much more than a Baker.
We truly don't have a name for it yet here on Earth.

.

Contents

Preface

Who of us, prior to our first chemistry class, would have imagined that a tasty white condiment like table salt is made from equal parts of an explosive gray metal (sodium) and a poisonous green gas (chlorine)? Yet the chemists tell us it is so; they can even make it before our very eyes. And that is only the beginning of the wonder. Who would have thought that all the multifarious materials we find around us are made up of a mere 100 different kinds of atoms in various arrangements?

There is a similar surprise lurking in the less familiar study of languages, I claim. Some 6,000 languages are spoken in the world today, and for the most part they seem very different from each other. Mohawk, for example, is quite a different thing from Japanese or English or Welsh or Swahili or Navajo or Warlpiri or Hixkaryana. Nevertheless, linguists are discovering that the differences among these languages are created by a small number of discrete factors, called parameters. These parameters combine and interact with each other in interesting ways to create the wide variety of languages that we can observe around us. Even though every sentence of Mohawk is different in its structure from corresponding sentences in English, and every sentence of English is different in its structure from corresponding sentences in Japanese, the "formulas" for making sentences in these three languages differ in only one factor each. In this respect, parameters can play the same foundational role in scientific theories of linguistic diversity that atoms play in chemistry. The purpose of this book is to give a sense of these fascinating discoveries. Along the way I explain precisely what a language is, what parameters are, and how a small change in parameters can create a large change in languages. I show some of the elegant and surprising similarities that underlie the obvious differences among languages. And I explore curious trivia about obscure South American languages that you can pass on to your friends at parties.

Moreover, the parametric theory of language diversity implies that the day is coming when linguists should be able to produce a complete list of the parameters that define all human languages, just as Dmitry Mendeleyev produced a table of the elements that define all physical substances. By the end of this book, I give a preliminary sketch of what such a periodic table of languages might look like.

This new view of how languages differ has significant implications for how we look at the relationship of language to thought and culture. Some believe that all languages are alike, and thus they attach no significance to linguistic diversity. Others believe that languages are incomparably different and thus their speakers are incapable of truly understanding each other. In contrast to both these views, the parametric theory of language leads to the conclusion that languages are different but commensurable. Their sentences are dissimilar, but their underlying principles are the same. Language differences therefore do not represent radically different worldviews or adaptations to different environments, but neither are they trivial or inconsequential. Different languages offer us slightly different perspectives, which can be compared and correlated to our mutual enrichment. As such, language differences can be seen as being part and parcel of the abilities to understand and interact with the world that make us persons.

Many people whose names are not on the title page and who are not likely to get a dime of the royalties have played a role in bringing this book into being. I hope it will be some comfort to them to see their names appear here, with my heartfelt thanks.

I particularly thank all the people who read the manuscript at one stage or another and told me what they did not understand (and in some places what I did not understand). There are many fewer such sections because of them. They include the linguists Veneeta Dayal, Ken Safir, and Roger Schwarzschild; chemist and fellow lover of analogies Jeff Keillor; biologist and fellow traveler Kyle Vogan; theoretical computer scientist and fellow McCloskey student David

Williamson; and representatives of the general public Jean Baker, Linda Baker, Dennis Cahill, and Tammy Knorr.

I also wish to thank David Pestesky, who got me started on the project of figuring out how to present linguistic results in an interesting way to a wider audience, an effort that eventually led to my writing this book. He and the late Vicky Fromkin, together with Joseph Aoun and Guglielmo Cinque, offered extremely detailed and perceptive criticisms of the material that became Chapter 4. Lila Gleitman and Jerry Fodor encouraged early versions of this work and helped me to sharpen my understanding of the relationship among language, concepts, and culture through the seminar they led at Rutgers University. Noam Chomsky and Steven Pinker gave helpful encouragement and suggestions when I was putting together the overall plan of the book and deciding how to approach the topic. Zsuzsa Nagy helped track down references, check typological generalizations, and find material on obscure languages. Thanks also to my Rutgers students who were the willing guinea pigs for much of this material in my typology class.

I also thank the many speakers of other languages who have patiently shared their invaluable knowledge with me. They include: Grace Curotte, Frank and Carolee Jacobs (speakers of Mohawk), Uyi Stewart (speaker of Edo), Sam Mchombo (speaker of Chichewa), and Ahmadu Kawu (speaker of Nupe). I would have much less to offer the world if it were not for them.

I thank Bill Frucht, my editor at Basic Books. We may not agree about basketball teams, but he recognized the potential of this project early on and has been an essential sounding board for what is interesting and what is not. He also tightened up my prose greatly, pointing out to me that it is not ideal to begin four consecutive sentences with *however*.

Finally, I thank my family and the small group that meets Sunday nights in our home for supporting me throughout the whole process, keeping me sane and focused on the ultimate goal.

And as always S.D.G.—which for me is also a reminder to Still Do Good.

Mark C. Baker

1

The Code Talker Paradox

DEEP MYSTERIES OF LANGUAGE are illustrated by an incident that occurred in 1943, when the Japanese military was firmly entrenched around the Bismarck Archipelago. American pilots had nicknamed the harbor of Rabaul "Dead End" because so many of them were shot down by antiaircraft guns placed in the surrounding hills. It became apparent that the Japanese could easily decode Allied messages and thus were forewarned about the time and place of each attack.

The Marine Corps responded by calling in one of their most effective secret weapons: eleven Navajo Indians. These were members of the famous Code Talkers, whose native language was the one cipher the Japanese cryptographers were never able to break. The Navajos quickly provided secure communications, and the area was soon taken with minimal further losses. Such incidents were repeated throughout the Pacific theater in World War II. Years after the end of the war, a U.S. president commended the Navajo Code Talkers with the following words: "Their resourcefulness, tenacity, integrity and courage saved the lives of countless men and women and sped the realization of peace for war-torn lands." But it was not only their resourcefulness, tenacity, integrity, and courage that made possible their remarkable contribution: It was also their language.

This incident vividly illustrates the fundamental puzzle of linguistics. On the one hand, Navajo must be extremely different from Eng-

lish (and Japanese), or the men listening to the Code Talkers' trans-
missions would eventually have been able to figure out what they
were saying. On the other hand, Navajo must be extremely similar
to English (and Japanese), or the Code Talkers could not have transmitted with precision the messages formulated by their English-
speaking commanders. Navajo was effective as a code because it had
both of these properties. But this seems like a contradiction: How
can two languages be simultaneously so similar and so different?
This paradox has beset the comparative study of human languages
for centuries. Linguists are beginning to understand how the paradox
can be dissolved, making it possible for the first time to chart out
precisely the ways in which human languages can differ from one another and the ways in which they are all the same.

———————

Let us first consider more carefully the evidence that languages can
be radically different. The Japanese readily solved the various artificial codes dreamed up by Allied cryptographers. Translating a message from English to Navajo evidently involves transforming it in
ways that are more far-reaching than could be imagined by the most
clever engineers or mathematicians of that era. This seems more remarkable if one knows something about the codes in use in World
War II, which were markedly more sophisticated than any used before that time. In this respect, an ordinary human language goes far
beyond the bounds of what can reasonably be called a code. If the
differences between Navajo and English were only a matter of re-
placing words like *man* with Navajo-sounding vocabulary like
hastiin, or putting the words in a slightly different order, decoding
Navajo would not have been so difficult. It is clear that the characteristics one might expect to see emphasized in the first few pages of
a grammar book barely scratch the surface of the complexity of a
truly foreign language.

Other signs of the complexity and diversity of human languages
are closer to our everyday experience. Consider, for example, your
personal computer. It is vastly smaller and more powerful than any-

thing the inventors of the computer imagined back in the 1950s. Nevertheless, it falls far short of the early computer scientists' expectations in its ability to speak English. Since the beginning of the computer age, founders of artificial intelligence such as Alan Turing and Marvin Minsky have foreseen a time in which people and computers would interact in a natural human language, just as two people might talk to each other on a telephone. This expectation was communicated vividly to the world at large through the 1968 movie *2001: A Space Odyssey,* in which the computer HAL understood and spoke grammatically perfect (if somewhat condescending) English. Indeed, natural language was not even considered one of the "hard" problems of computer engineering in the 1960s; the academic leaders thought that it would more or less take care of itself once people got around to it. Thirty-five years and billions of research dollars later, their confidence has proved unwarranted. It is now 2001, and though HAL's switches and indicator lights look hopelessly out-of-date, his language skills are still in the indefinite future. Progress is being made: We only recently achieved the pleasure of listening to weather reports and phone solicitations generated by computers. But computer-generated speech still sounds quite strange, and one would not mistake it for the human-generated variety for long. Moreover, these systems are incapable of improvising away from their set scripts concerning barometric pressures and the advantages of a new vacuum cleaner.

This poor record contrasts with scientists' much greater success in programming computers to play chess. Another of HAL's accomplishments in *2001* was beating the human crew members at chess— a prediction that has turned out to be entirely realistic. We usually think of chess as a quintessentially intellectual activity that can be mastered only by the best and brightest. Any ordinary person, in contrast, can talk your ear off in understandable English without necessarily being regarded as intelligent for doing so. Yet although computer programs can now beat the best chess players in the world, no artificial system exists that can match an average five-year-old at speaking and understanding English. The ability to speak and understand a human language is thus much more complex in objective terms than the tasks we usually consider to require great intelligence. We simply tend to take language for granted because it comes so quickly and automatically to us. Just as Navajo proved harder than other codes during World War II, so English proves harder than the Nimzowitsch variation of the French defense in chess.

The experience of computer science confirms not only that human languages are extremely complex but that they differ in their complexities. Another major goal of artificial intelligence since the 1960s has been machine translation—the creation of systems that will take a text in one language and render the same text in another language. In this domain the ideal is set not by HAL but by *Star Trek:* All crew members have a "universal translator" implanted in their ears that miraculously transforms the very first alien sentence it hears into perfect English. Again, real machine translation projects have proven more difficult. Some programs can take on tasks like converting the English abstracts of engineering articles into Japanese or providing a working draft of a historical text from German in English or translating a page on the World Wide Web. But the products of these systems are very rough and used only in situations where an imperfect aid is desired. Indeed, sometimes they make embarrassingly funny mistakes. Harvey Newquist reports an apocryphal story about an early English-Russian system that translated the biblical quotation "The spirit is willing, but the flesh is weak" into Russian as "The vodka is strong, but the meat is spoiled." Performance has improved since the 1960s, but not as much as one might imagine. Here is a quotation from the biblical book of Ecclesiastes (9:11 RSV):

> Again I saw that under the sun the race is not to the swift, nor the battle to the strong, nor bread to the wise, nor riches to the intelligent, nor favor to the men of skill, but time and chance happen to them all.

Here is the same passage after it has been translated into Russian and back again by a randomly selected Web-based machine translation program:

Afresh I beheld such under the sun the race am no near the fast, no bread near the profound, no affluence near the clever, no favor near the body *[ÿíêå]*) against art, alone fardel ampersand accident become near their all.

It is fair to say that something is lost in this translation. My friend who translates banking documents from English to French need not look for other work just'yet.

Moreover, the best systems so far work on an ad hoc basis, taking advantage of whatever special properties they can find in the two languages they deal with. When good systems are finally up and running for Russian-English translations, they certainly will not be any good at Navajo or Swahili or Turkish. Adapting the program to these languages will not be a matter of adjusting a few settings. Rather, programmers will have to start almost from scratch to create comparable systems for these languages.

Lest we be tempted to look down on computers and Japanese cryptographers, we should take stock of our own experiences in learning foreign languages. Having been deeply moved by stories of John Henry and the steam engine as a youth, I have a fondness for stories about what machines cannot do. American patriots might feel smug about how easily Japanese intelligence was duped by the clever Marine Corps. But what about us? With the right kind of exposure to a language and plenty of hard work, adults can achieve a reasonable degree of fluency in a new language. But very few, even after years of living in another country, ever learn a foreign language so well that they could pass as a native speaker. Most of us never even make it up to the level of full fluency. For example, I took five years of Spanish in an American high school. I aced most of the grammar tests, but no Spaniard will ever mistake me for one of his own. And even after living in Montreal for twelve years I find it embarrassingly difficult to follow a hockey game in French without visual aids. And French, Spanish, and English are all closely related by global standards: For a native English speaker, learning Chinese or Arabic or Turkish is harder still. Only a handful of white people have ever learned Navajo with any degree of fluency. Americans should be thankful that the Japanese empire had no aboriginal languages of its own to press into service.

———————

What was it about Navajo that made it so difficult to decode? As far as I know, no one has investigated what strategies the cryptographers tried and precisely why those strategies failed. But if one knows a little bit about Navajo, it is not hard to guess some of the reasons. Much of the difficulty in coming to grips with an unfamiliar language is that there are many layers of difference. Each layer might be understandable enough in its own terms, but the differences magnify each other until the total effect is overwhelming.

First, of course, there is the fact that the Navajos (to adapt an old Steve Martin joke about French) have a different word for everything. When an English speaker would say *girl,* a Navajo speaker would say *at'ééd;* for English *boy,* a Navajo would use *ashkii;* for English *horse,* Navajo has *łįį',* and so on. Moreover, the things that English has specific words for and the things that Navajo has specific words for do not always match perfectly. For example, English is unusually rich in words for various modes of thought and feeling. In English, we can believe, know, wonder, opine, suppose, assume, presume, surmise, consider, maintain, and reckon. We can be furious, irritated, incensed, indignant, irate, mad, wrathful, cross, upset, infuriated, or enraged. In many other languages, this domain of meaning is covered with only two words, 'to think' and 'to be angry.' Navajo, for its part, has at least ten different verbs for different kinds of carrying, which depend on the shape and physical properties of the thing being carried: *'Aah* means to carry a solid roundish object, such as a ball, a rock, or a bottle; *kaah* means to carry an open container with its contents, such as a pot of soup or a basket of fruit; *lé* means to carry a slender flexible object like a belt, a snake, or a rope; and so on. For this reason, finding the right words to use in a translation to or from Navajo involves much more than simply substituting one string of letters for another. The "Replace All" command on your word processor will never be able to do it properly.

Second, there are important differences in the sounds that make up words in Navajo. As you surely noticed, the Navajo words listed above contain some strange-looking symbols: *l*'s with bars through them, vowels with accents over them and hooks under them, apostrophes in the middle of words. This reflects that the Navajo language is built around a different set of basic sounds than English is. For example, the *ł* stands for a sound that is rather like that of the English *l* but made "whispering," without vibration of the vocal cords. (The same sound is indicated by the double *ll* in Welsh words like *Lloyd*.) Hooks under the vowels indicate that these sounds are pronounced nasally, with air passing through the nose as well as the mouth. The Navajo word *sá* 'old age' is pronounced much like the French word *sans* 'without.' To complicate matters further, the specific qualities of these sounds adjust in complex ways to the sounds around them. For example, Navajo has a prefix *bi-* that attaches to nouns and means 'his' or 'hers.' Thus, *gah* means 'rabbit' and *bigah* means 'his/her rabbit.' When this prefix attaches to certain words that begin with *s*, that *s* changes to a *z* sound. Thus, *séí* means 'sand,' but *bizéí* means 'his/her sand.' In the same context, the whispered *ł* sound undergoes a similar change to become plain *l*: *łíį́'* is 'horse,' and *bilį́į́'* is 'his/her horse.' These differences in sound are significant because it is notoriously difficult for people to recognize sounds that do not exist in their own languages. Japanese speakers have a terrible time distinguishing English *l* versus *r*, whereas English speakers have trouble recognizing the four different *t* sounds in Hindi. When added together, the many sound differences give Navajo speech a very distinctive, almost unearthly quality that speakers of Eurasian languages find difficult to grasp or remember.

Words in Navajo also change their form depending on their context in various ways. One of the most remarkable aspects of Navajo is its system of prefixes. Indeed, simple, invariant words are rare in Navajo. We just saw that a prefix can attach to a noun to show that the noun is possessed. The prefix system for verbs is even more elaborate. There are between 100 and 200 different prefixes that attach to Navajo verb stems, depending on the exact analysis. Even the simplest verbs in Navajo must take at least three prefixes, five or six are common, and a verb can have up to ten or twelve at one time. The total number of forms a Navajo verb can take is staggering. Nor can the language learner afford simply to overlook these prefixes and hope for the best. The subject of the sentence, for example, is often hidden inside the prefixes. 'The girl is crying' in Navajo is a fairly normal-looking combination of the word for 'girl' and (one form of) the word for 'cry.'

> At'ééd   yicha.
> Girl   crying

But 'I am crying' is expressed by a verb standing alone. That I am crying, not you or they, is expressed by a prefix *sh-* found before the verb root but after other prefixes.

> Yishcha.   (yi + sh + cha)
> 'I am crying.'

(This ability of some languages to express subjects as changes on the verb plays a major role in my discussion in Chapter 4.) Other Navajo prefixes elaborate on the basic meaning of the verb root in intricate ways. For example, the simple root *dlaad,* meaning 'to tear,' combines with six different prefixes to make the following word, meaning 'I am again plowing.'

> Ninááhwiishdlaad.   (ni + náá + ho + hi + sh + ł + dlaad)
> 'I am again plowing.'

These aspects of Navajo pose a major challenge to that great institution of Western civilization, the dictionary. Since Navajo has so many prefixes, the primary lexical meaning is rarely carried in the first part of a word. Thus, the basic idea of listing words in alphabetical order is not so practical for this language. Looking up a word in Navajo requires first identifying the prefixes, undoing the sound changes that they cause, and deleting them. Only then can one find the basic root of the word and calculate the changes in meaning caused by the prefixes. Dictionaries of Navajo do exist, but successfully using one is a major intellectual achievement—the way you prove you have mastered Navajo, not the way you learn it.

Navajo also has complexities at the level of syntax, how its words are put together to make phrases and sentences. The simplest subject-verb combinations, like 'the girl is crying' shown above, look innocent enough: The subject noun phrase comes first, as in English, and the verb that expresses the predicate comes second. But differences appear in more complex sentences. For example, consider a transitive sentence, one that contains a direct object noun phrase as well as a subject. In Navajo the direct object always comes before the verb, not after it as in English:

Ashkii   at'ééd   yiyiiłtsą.
Boy   girl   saw
'The boy saw the girl.'

*Ashkii   yiyiiłtsą   at'ééd.
Boy   saw   girl

(Linguists put an asterisk in front of an example to show that the way of combining words is impossible in the language under discussion. I use this convention frequently.) Other phrases have a distinctive word order, too. Whereas in English one says 'change into your clothes,' the Navajo would say the equivalent of 'clothes into change.' Whereas in English one says 'John believes that he is lying,' in Navajo one would say the equivalent of 'John he lying-is believes.' In fact, there is a systematic pattern to these Navajo word orders, a topic I discuss in detail in Chapter 3. But systematic or no, it is confusing to an English speaker.

If the Japanese cryptographers had got this far, they might have breathed a sigh of relief at this point, because these word order patterns are actually the same as in Japanese. But their newfound confidence would have evaporated when they came across sentences like this:

Ashkii   at'ééd   biiłstą́.
Boy   girl   saw

This sentence has almost the same words arranged in the same order as the one we saw above, and so we might reasonably guess that it has the same meaning: 'The boy saw the girl.' In fact, this sentence means the opposite, that the girl saw the boy. The crucial hint is once again in the prefixes attached to the verb: Here the verb starts with *bi-*, whereas the verb in the previous sentence started with *yi-*. This small difference indicates a large difference in sentence structure. *Bi-* tells the Navajo speaker roughly that the direct object of the sentence comes before the subject, rather than the other way around. Nor is it always easy to find these prefixes: Like any others in Navajo, they can be buried under other prefixes and disguised by sound changes. Furthermore, this option of choosing a sentence with *yi-* or a sentence with *bi-* is used in a culturally specific way in Navajo. The noun phrase that refers to a higher being always comes before a noun phrase that refers to a lower being, regardless of which is the subject and which is the object. (In the Navajo conception, humans count as higher than large and intelligent animals, which count as higher than smaller animals, which in turn count as higher than plants and inanimate objects.) Sentence structure and word structure thus are interdependent in Navajo, and both are influenced by the distinctive Navajo typology of creatures, which puts hawks below wildcats but on the same level as foxes. Not only is Navajo different from English at many levels, from single sounds to the arrangement of words in sentences, but those different levels interact with each other in various ways. There is a combinatorial explosion of difference, and it seems as if one cannot understand anything until one understands everything. The poor Japanese cryptographers didn't stand a chance.

These striking differences between Navajo and English illustrate only one side of the fundamental puzzle of language. The other half of the Code Talker paradox is that Navajo and English are so similar.

The Code Talkers bore witness to this similarity by their ability to translate messages back and forth between Navajo and English. Originally, the U.S. Marine officials doubted whether this would be possible and were reluctant to pursue the project. Pilot studies, however, proved that the Indians could accomplish the task with great accuracy. You are probably familiar with the game of Telephone, in which a message gets garbled beyond recognition as it is whispered from one child to another. Encoding the message into Navajo and decoding it back into English did not significantly increase this garbling. On the contrary, officials were pleasantly surprised at how well even precise technical information was preserved. More than that, the Code Talkers were *fast*. They could translate a message to and from Navajo almost instantaneously, in a fraction of the time it took to encrypt it by conventional means. This made them invaluable in battle situations, in which circumstances could change rapidly, and seconds counted.

The Code Talkers' performance tells against extreme versions of one common view about language differences. Some people believe languages are so different because they are reflections of human cultures that have developed over time in diverse environments and in relative isolation. Different languages thus represent incommensurable ways of thinking about the world. If that were so, the early suspicions of the marine brass should have been borne out: The Code Talkers' translations should have been laborious and inaccurate, if they were possible at all. Certainly nothing in the Navajos' pastoral lifestyle in the Arizona deserts would have helped them conceive of high-tech warfare in the jungles of the Pacific Islands. Yet they did their task with remarkably little training. Apparently English and Navajo—or any other two languages—are not products of incommensurable worldviews after all. They must have some accessible common denominator.

Again, this conclusion is reinforced by our mundane experience, which tells us that translation between two languages is a commonplace experience. Although the limitations of machine translation projects show that translation is hard, the everyday successes of professional human translators prove that it is possible. It is stylish to disparage translations, saying that they are (like mistresses, I am told) either ugly or faithless. But these complaints are generally raised in a literary context. When translating Dante or Shakespeare, one wants to preserve not only the literal meaning of the text but also its meter and rhyme scheme, its connotations and cultural allusions, its puns and wordplay, and the ingenious resonances between sound and meaning that give rise to great poetry. That task is often impossible, simply because there are too many conflicting demands. Away from the domain of high art, however, the idea of translation becomes much less problematic. My friend translates internal banking documents from English to French, and she does a good job. At least the money goes to the right places. To take another example, probably more non-European languages have been first learned by Westerners out of a desire to translate the Bible into the local tongue than for any other reason. The people on both ends of these spiritually motivated translation projects have often felt that something real was being done.

Finally, consider Philip Johnston, the man who first conceived of the Code Talkers project. He was one of the small handful of white people who could speak Navajo fluently. He managed this without being ethnically Navajo himself; nor did he have the education and resources of the Japanese cryptographers when he broke the Navajo code. What was his secret? Simply this: as the son of Christian missionaries, he grew up playing with Navajo-speaking children and so learned the language as a child. Although it is very rare for an adult to learn a foreign language perfectly, this achievement is commonplace for a child raised in the right environment. This phenomenon is seen over and over again in this age of migrations. For example, many East Asian immigrants in New Jersey make their way through life with labored and broken English, but their children's English is indistinguishable from that of a Smith or a Jones.

The more one thinks about this commonplace achievement, the more remarkable it seems. A language is one of the most complex systems of knowledge that a human being ever acquires. Learning one far outstrips such feats as memorizing the capitals of every nation of the world or learning the rules for proper capitalization. Yet children, without graduate training, government funding, or even adequate secretarial help, manage in five years what computer scientists and linguists have been wrestling with unsuccessfully for fifty. And unlike academics, the children don't even have the advantage of already knowing a human language as a basis for comparison when they set out to learn the one spoken around them. Nor are we talking about a few prodigies here, but about every normal, healthy child.

How is this amazing feat accomplished? Linguists and other cognitive scientists conclude, not without some envy, that the children must have a huge head start. By their very nature, children seem to be specially equipped for language learning. No one yet knows exactly what this special equipment consists of. It probably involves knowledge of what human languages are like and of what kinds of sounds and structures they might contain, together with strategies for recognizing those sounds and structures. Linguists call this innate head start "universal grammar," an idea popularized by Steven Pinker as the "language instinct."

This has implications for our questions about linguistic diversity because whatever this universal grammar is, it must equip children to learn any of the myriad of possible human languages. Prior knowledge of the special properties of English might help a human child growing up in Spooner, Wisconsin, but it would only hinder a child learning Mapuche in Junín de los Andes, Argentina. It is conceivable that children could begin life with prior knowledge not just of one human language but of some 10,000 different ones. Then their task would not be to *learn* the complexities of the language being spoken around them but merely to *recognize* which language it is and forget the rest—presumably an easier project. But this scenario is hardly plausible. The idea of an inborn knowledge of grammar is already incredible enough; to have many such inborn systems existing side by side would be uneconomical in the extreme. Nobody believes that we all have innate knowledge of Ancient Akkadian. The expectations children bring to language learning must be substantial enough to help them master the complexities of language with relative ease yet general enough to be applicable to any human language they come into contact with. This implies that all human languages must be more similar than they appear. Since all are equally within the grasp of a healthy human child to learn through ordinary, informal exposure, they must, with all their distinguishing intricacies, be fundamentally commensurable. This conceptual argument from language learning, even more than the phenomenon of intertranslatability, convinces linguists that at some level all languages must be the same.

———

Can even languages as different as English and Navajo be cut from the same cloth? Yes. We have already surveyed some of the salient features of Navajo, focusing on the differences. But even in that discussion there were signs of similarity as well. Some of the basic sounds of Navajo are different from English, but others are the same: Both languages have *p, t, m, n, w, y, a, e, i,* and *o.* Even the sounds that are different, such as *l,* can be described as being like an English sound with one feature of the pronunciation changed—in this case, the vibrating of the vocal cords. In both languages sounds adjust to their environments in comparable ways. For example, *s*'s turn into *z*'s in English, too, as you can see by listening carefully to yourself pronounce the word *dogs* (one actually says "dogz"). Like Navajo, English verbs also inflect to show tense and subject; compare the verb in *They walk* with *They walked* and *Chris walks.* It is true that Modern English has only a few of these inflections, and they are suffixes rather than prefixes. Its inflections are like a Ford Escort compared to Navajo's Lamborgini. Nevertheless, it has the rudiments of the kind of system that is so highly developed in Navajo. Direct objects come before the verb in Navajo and after the verb in English, but both languages have direct objects and verbs. More than that, they are similar in that in basic sentences the direct object appears right next to the verb, whereas the subject need not. In Navajo the object comes between the subject and the verb, whereas in English adverbs and tense auxiliaries can separate the two (e.g., *Chris will soon find a quarter*). This similarity foreshadows a very significant universal feature of human language important in subsequent chapters.

Perhaps the single most distinctive property of Navajo is that its word order changes when the verb contains the prefix *bi-* rather than *yi-*.

Ashkii	at'ééd	biiłstą́.		At'ééd	ashkii	yiyiiłtsą́.
Boy	girl	saw		Girl	boy	saw
'The girl saw the boy.'				'The girl saw the boy.'		

This is indeed a capstone of Navajo grammar and an endless source of fascination and controversy among the people who study this language and its relatives. Nevertheless, English does have something somewhat similar. In English, too, it is possible to permute the basic phrases of a sentence together with a change in the form of the verb. The English permutation is known as the passive voice:

The girl saw the boy.
The boy was seen by the girl.

It would be underestimating the Navajo *yi-/bi-* alternation to say that it is exactly like the English passive. There are as many differences as similarities. For example, the agent of the passive verb in English, marked by the preposition *by,* can be left out of the sentence altogether:

The boy was last seen at the market.

This sentence means that somebody saw the boy but does not say who. In contrast, the repositioned subject in Navajo *bi-* sentences has no special trappings and cannot be omitted:

*Ashkii   biiłstą́.
Boy   saw
Bad as: 'Somebody saw the boy.' (OK only as: 'The boy saw him/her.')

Yet even here, where the languages seem most different, there is a suggestive point of similarity.

———————

So is Navajo very different from English? Yes. Or are the languages similar? Also yes. This is the Code Talker paradox again, seen closer up. Just when linguists convince themselves that two languages are incomparable, they come across a striking analogy between them. Then again, just when linguists think they have discovered a single theory that works for both, they are ambushed by a surprising distinction between the two. These issues arise whenever one compares two languages that are not close neighbors geographically and historically. The fundamental challenge of comparative linguistics is to find a way of doing justice to both the similarities and the differences without contradiction, without empty compromise, and without sacrificing one truth to the other. Linguists have recently found a new way of thinking about these problems that has the promise of finally meeting this challenge.

Our glance at Navajo makes it clear that the paradoxes about similarity and difference arise at many levels, from individual sounds to the interrelationships of whole sentences. It would be bewildering, however, to try to focus equally on all of these levels in a book of this nature. Nor is it necessary, because the conceptual issues seem to be essentially the same at each level. Therefore, this book concentrates primarily on syntax, together with those aspects of the form of words that are related to syntactic issues. Thus, I lead you into more detail about topics such as word order, pronoun omission, and the *yi-/bi-* alternation, while putting aside questions of whispered *l* sounds and how they change when a prefix is added. This latter domain—called phonology—is no less instructive. But syntax is a large subfield of linguistics and perhaps the one most closely connected with questions about thought and culture. It is also the area that first inspired the conceptual innovations I go on to discuss. Finally, it happens to be the area where I am best qualified to act as a guide.

Before beginning the exploration in earnest, it is fair to pause and ask the immortal question: "Why should I care?"

For better or worse, people do care passionately about languages and the distinctions among them. There may be areas of the United States where it is possible to overlook this fact, simply because most people are monolingual. But such areas are uncommon in the world as a whole. The impression that language differences don't matter can be cured by talking to the Mohawk Indians who feel their language threatened by the French Québecois majority around them, who in turn feel their language threatened by the English Canadian majority around them. Or one can go to certain African countries where the language spoken by the chair of the Linguistics Department at the country's flagship university is a source of political conflict and even violence. Or one can visit parts of the United States where English-only legislation is controversial and divisive.

The reason for this passion, of course, is that language is intimately tied to culture, and the question of the commensurability of languages is related to the question of the commensurability of cultures. Many scholars and many ordinary people believe intuitively that much of our higher thought processes as well as our culture is intertwined with our language. Just what these interrelationships are is very difficult to tease apart, but there is no doubt that much of our conscious thought and cultural knowledge is framed in the medium of language. The same tensions between similarity and difference arise in these other domains. Are two cultures basically variations on a common theme, or do they represent deeply different ways of perceiving and thinking about the world? If the latter, then are they incommensurable and hence incapable of truly understanding each other? These questions are the cultural equivalent of the Code Talker paradox, and again the evidence points both ways, with clear disciplinary biases. Anthropologists thinking about culture tend to emphasize difference, whereas cognitive psychologists thinking about thought tend to emphasize similarity. One of the charms of linguistics is that it stands at the crossroads of these two intellectual traditions, where neither emphasis can easily be ignored.

Of the human triumvirate of culture, thought, and language, language is the most accessible to rigorous intellectual study. Your tape recorder can give you an objective portrayal of what I say but not of my thoughts or my cultural understanding. If there is hope of solving the Code Talker paradox and thus achieving a more mature understanding of linguistic diversity, we will gain a useful metaphor for thinking about cultural and mental diversity as well. This could powerfully influence how we understand human nature and how we cope with life in a pluralistic world—whether we dread it, are defeated by it, or are able to relish it.

2

The Discovery
of Atoms

THE KEY TO RESOLVING A PARADOX often lies in the imagination. In a paradox where some experience points to one conclusion and other experience seems to point to its opposite, what is needed is not simply more experience. Truth is not a democratic matter to be decided by simple majorities. Rather, what is needed is some new idea that can widen the space of hypotheses. Then the conflicting evidence that seemed to lead to contradictory conclusions can be seen to converge on this new possibility.

So it is with the Code Talker paradox. To do justice both to the facts that show that languages are very different from one another and those that show they are similar, we need a new concept. Linguists call their new concept the *parameter*.

Since linguistics is a relatively young and unfamiliar science, I introduce the history of the idea of a parameter by comparing linguistics to an older and more familiar discipline, namely, chemistry. Just as chemistry learned to address the paradoxical properties of physical substances and their transformations in terms of atoms, so linguistics addresses the paradoxical properties of languages in terms of parameters. We may think of parameters as the atoms of linguistic diversity.

In their subject matter, methodology, and historical development, chemistry and linguistics are not even distantly related. Chemistry is unquestionably a physical science; linguistics is generally classified as a social science or in the humanities. The two departments are almost never housed in the same building at the universities. Nor is the gap between chemistry and linguistics bridged by any of the compound fields that have become common in academia: There are no "chemolinguists" or "linguochemists" with joint appointments in both departments. What two objects of study could be more different than Navajo and nitroglycerin?

Nevertheless, there are some abstract similarities between the basic phenomena that chemistry and linguistics are concerned with. Like linguistics, the foundational paradoxes of chemistry concern the tensions between sameness and difference, between stability and change. Indeed, the root of the word *chemistry* comes from the Greek word *khēmeia,* meaning 'transmutation.' Chemistry was originally about how a substance with one set of properties (say, lead) could be changed into one with very different properties (with luck, gold). This is analogous to the question of how the Code Talkers could take a message in a language with one set of characteristics (English) and change it into a language with very different characteristics (Navajo). Chemistry is about *transmutation;* linguistics is about *translation.* At an abstract level they both encounter the same paradox: How can radically different entities yet be sufficiently alike that one may transform into another?

Thought about the chemical version of this paradox is very old, dating back to the pre-Socratic natural philosophers of ancient Greece. On the one hand, experience combined with reason told them that a pure substance left to itself would always remain pure: It could not come into existence, pass out of existence, or change its fundamental mode of existence. Pure water left undisturbed in a covered cup always remains pure water. On the other hand, they also observed that the properties of things did change over time. Wood burns, producing smoke, ashes, and flame, each of which has very different physical characteristics from the original wood. It is unwise, for example, to build a house out of ashes, smoke, or flames. And they wrestled with the tension between these two basic observations. How could wood be so different from ashes and smoke and yet similar enough to be trans-formable from one to the other? This was their version of the Code Talker paradox incarnated in the physical world.

Eventually the ancient Greeks imagined a possible answer. They reasoned that most of the substances around them must be mixtures of more basic substances, which were the true elements of the uni-verse. A basic substance such as water (they thought) truly could not change; it would be genuinely eternal and immutable within the scope of this creation. But basic substances could combine in various ratios and arrangements to form other, derived substances. The prop-erties of a derived substance would be some kind of blend of the properties of the elements that make it up. In this way, the derived substance could be quite different from any elementary substance. This can in principle explain why so many different types of matter can arise from a limited number of basic elements (four, according to Empedocles). It also explains why most substances undergo change. Their properties change if the proportions of basic elements in them change. For example, wood is a derived substance made out of a cer-tain amount of ashes (earth), some smoke (air), and a healthy dash of fire, with perhaps a hint of water thrown in to taste. Burning is not a fundamental change in substance but a coming apart of three basic substances that were temporarily mixed. In this way, sub-stances with very different properties may still be considered com-mensurable, because they are made up of the same elementary building blocks.

Toward the end of this intellectual tradition, Democritus further re-alized that for this view to work, the elements themselves must come in discrete, very small pieces, which he called atoms (*átomos* means 'undivided' in Greek). Atoms must be tiny, because everyday-sized chunks of matter are always divisible. For example, wood does not contain visible pieces of ash, smoke, and fire, so the atoms must be too small to see. Water in an open glass will gradually, imperceptibly disappear. If no element can come into or out of existence or change its essential nature, then this must be because tiny pieces of water fly off one at a time. Still, these basic chunks of matter could not be infinitely small, or else they couldn't combine into visible chunks. The sum $0 + 0 + 0 + \ldots + 0$ is 0 to infinity, and we know that there is more in the world than zero. Thus, the Greeks' best answer to the basic paradoxes of transmutation was atoms. And although they were wrong in all the details, they were right in the basic conception.

Maybe atoms can be answers in linguistics as well. Since the paradoxes of translation are partially analogous to the paradoxes of transmutation, it is reasonable to ask if they can be resolved in a similar way. Noam Chomsky, the Democritus of modern linguistics, has argued that this is the case (although without this chemical analogy in mind). Suppose languages are also composites of a finite number of more elementary factors, which Chomsky calls *parameters*. Different combinations of these parameters would yield the different languages we observe in the world. Indeed, a relatively small number of parameters might underlie the large number of possible human languages. If the parameters are also like chemical elements in that they interact with each other in complex and interesting ways, then the properties of the resulting languages might show striking variation. Chomsky writes:

> What we expect to find, then, is a highly structured theory of UG [universal grammar] based on a number of fundamental principles . . . with parameters that have to be fixed by experience. If these parameters are embedded in a theory of UG that is sufficiently rich in structure, then the languages that are determined by fixing their values one way or another will appear to be quite diverse, since the consequences of one set of choices may be very different from the consequences of another set; yet at the same time, limited evidence, just sufficient to fix the parameters of UG, will determine a grammar that may be very intricate . . . . What seems to me particularly exciting about the present period in linguistic research is that we can begin to see the glimmerings of what such a theory might be like.

On the face of it, it might seem astonishing that a language like Navajo is made up of the same basic parts as English or Japanese. But this is no more astonishing than the chemical fact that a tasty white condiment (table salt) can be made up of the same basic parts as an explosive gray metal (sodium) and a poisonous green gas (chlorine).

Such a theory can also do justice to the evidence that languages are commensurable. If Navajo and English are built from the same basic elements, with only the proportions and arrangements being different, then it is less surprising that there would be reliable algorithms for transforming one into the other. These algorithms are what the Code Talkers mastered and machine translation projects seek. Their function would be comparable to "translating" graphite into a diamond by rearranging the same chemical stuff.

The idea that all languages are combinations of a finite number of basic parameters also sheds light on the more fundamental paradox of language acquisition. If there are such things as parameters, then children can come to the task of language learning preloaded with a knowledge of these basic parameters and an ability to deduce the chemistry of their interactions. These parameters and regulating principles are what Chomsky calls "universal grammar." Children with an innate knowledge of this universal grammar would learn a language like English or Japanese or Navajo simply by establishing which of the parameters are present in that particular language and in which ratios or arrangements. Although not a trivial task, this is enormously easier than discovering linguistic structure from scratch. This can explain how children can have a big head start on the task of language learning that is equally applicable to learning English or Navajo or whatever other language their playmates happen to speak.

Finally, a theory based on parameters could explain the fact that languages also have a kind of mutability. Over time, a language like English can evolve to become grammatically more like Navajo and vice versa. (I mention specific examples in Chapters 4 and 7.) The change could happen if the parameters that define English changed one at a time until eventually they resembled the "linguistic formula" for Navajo. Since there are only a finite number of parameters and ways of combining them, it is not astonishing that structurally similar languages should develop independently in different parts of the world, just as it is not surprising that methane should be formed independently on Jupiter and on Earth. Thus, a linguistic theory of parameters holds the promise of explaining any number of puzzles.

There is more to science than thought experiments and more to chemistry than Democritus. In the same way, there is more to linguistics than Chomsky. The best ideas must ultimately be tested against facts.

After Democritus, the next crucial contribution to chemistry came from the alchemists of the Middle Ages and the Renaissance, who were very different in temperament from the Greek natural philosophers. They thought less and did more. They didn't worry about how change was possible in principle. Instead, they strove to control change in practice so that they could make lead into gold, find the universal solvent, and live forever. They failed dramatically at these goals, but they gained practical experience with actual chemical reactions and developed the experimental procedures that eventually led to the discovery and isolation of the true chemical elements as alchemy gave way to chemistry. For however right the Greeks were about the existence of atoms and elements, they were dead wrong about what those elements were. Their favorite candidates—earth, water, fire, and air—turn out to be extremely heterogeneous substances chemically, about as far away from true elements as one could imagine.

If Chomsky is the Democritus of linguistics, then the anthropological linguists and the American structuralists have been its alchemists. Nineteenth-century European scholars, following in the wake of empire builders and missionaries around the world, began to get their first real impression of the true diversity of human language. This exposure increased markedly in the early twentieth century, as linguists who had come to America made a practice of studying American languages. Pioneers such as the anthropologist

Franz Boas, his student Edward Sapir, and their contemporary Leonard Bloomfield learned the intricacies of languages such as Lakhota, Paiute, and Menominee along with French, German, and Latin, and they taught others to do the same. Boas and Sapir in particular were driven by an urgent desire to obtain information about Native American languages before they disappeared and by a belief that studying these languages in their own terms (not through the lens of European grammatical concepts) would give deep insight into their cultures as well. Their work was particularly significant because Europe happens to be relatively homogenous linguistically. All European languages come from only two major families: the Indo-European family (the majority of them) and the Finno-Ugric languages (Finnish, Estonian, and Hungarian), together with one notable isolate, Basque. As such, their grammatical properties are rather similar. In contrast, Native North America contained ten or twelve major language families, each less closely related to the others than English is to Russian or Hindi. Those families in turn each included dozens of mutually unintelligible languages. The mountains of California alone had more linguistic diversity than all of Europe put together. Part of the legacy of Boas and his followers was to refute the myth that all North American languages were similar, characterized by a handful of evolutionarily "primitive" features. His generation also played a major role in developing the methodological infrastructure for investigating and describing non-Indo-European languages. Bloomfield not only credits Boas with providing masses of descriptive material but wrote that "Boas forged, almost single-handed, the tools of phonetic and structural description."

With this exposure, linguists came into contact with new evidence for the extreme differences among languages. They wrestled with exotic expressions like *inikwihlminih'isit* from Nootka and *wiitokuchumpunkurüganiyugwivantümü* from Paiute. Both these words are very complex; indeed, complexity of words is a distinctive property of many North American languages. The Nootka word is made up of five distinct parts: It is *inikw* 'fire' or 'burn,' plus *ihl* 'in the house,' plus *minih* 'plural,' plus *'is* 'small,' plus *it* 'past.' The

Paiute word contains no fewer than nine parts, working together to give the meaning 'they who are going to sit and cut up a black cow with a knife.'

Beyond their raw complexity, Nootka words seem to transcend the distinction between nouns and verbs that has always been central to European linguistics. *Inikwihlminih'isit* can be used equally well as a noun or a verb. As a noun, it is translated into English as 'the little fires that were once burning in the house'; as a verb, a suitable translation would be 'several small fires were burning in the house.' Apparently, Nootka grammar does not make use of this most basic European notion. Thus, Edward Sapir wrote in 1921 that "speech is a human activity that varies without assignable limit as we pass from social group to social group." It seemed that with language anything goes.

This great descriptive project continues to the present. Since the 1970s, Australian scholars have devoted enormous effort to characterizing their continent's aboriginal languages. For the most part, these languages do not have individual words that are as complex as those cited above from North American languages. But they have wonders all their own. For example, the words of a phrase in languages like English must appear next to each other. In a sentence like *These small children chased those big dogs,* we know that the children are small and the dogs are big because the word *small* appears next to *children* and the word *big* appears next to *dog.* Warlpiri and other Australian languages, however, have special features that allow their phrases to be split up and scattered across the sentence. This English sentence thus could be translated into Warlpiri in any of the following ways, among many others:

> These-su big-ob children-su chased those-ob small-su dogs-ob.
> These-su big-ob small-su chased children-su dogs-ob those-ob.
> Dogs-ob big-ob chased children-su small-su those-ob these-su.

Warlpiri's secret is that it puts certain endings, called *case markers,* on each word to indicate whether that word applies to the subject or the object. In my pretend-Warlpiri examples, I give a sense of this by putting SU (for *subject*) or OB (for *object*) at the end of each English word. Since information about which is which is encoded in this way, it is not necessary to group related words into phrases in Warlpiri the way it is in English. Rather, word order expresses other aspects of the message, such as what is important new information. If one knew only the modern European languages, one might falsely conclude that sentences like these were impossible.

This kind of work goes on today as more areas become accessible. In South America linguists have discovered languages that consistently have objects at the beginning of the sentence, a pattern that was once unheard of. The island of New Guinea has also been of special interest, because its high mountains and heavy rainfall have kept people relatively isolated from one another. As a result, New Guinea is home to 14 percent of the world's languages, virtually all of them unknown to linguists until recently. Linguists feel an increasing urgency to advance the task of describing and preserving the world's languages, as the forces of globalization, habitat destruction, and prejudice conspire to push more and more of them toward extinction. Just as a mature chemistry would not have been possible without the discovery of elements such as phosphorus and arsenic, which were unknown to the ancient Greeks, so a mature linguistics is not possible without knowledge of Nootka and Paiute and Warlpiri— and whatever rare elements may exist in the languages of the upper Amazon and the New Guinea highlands.

———

Once a critical mass of chemical elements had been found, chemists began to notice larger-scale patterns: Groups of elements have systematic similarities in their physical properties and chemical behavior. In the 1820s, Johann Wolfgang Döbereiner, an early leader in this project, found several groups of three elements, which he called "triads," that had similar chemical characteristics and whose atomic weights formed arithmetic progressions. For example, chlorine, bromine, and iodine form similar compounds, and the atomic weight of bromine (80) is (approximately) the average of the atomic weights of chlorine (35.5) and iodine (127). The existence of these triads pointed toward a systematic, law-governed relationship between atomic weight and chemical behavior. These patterns were essential stepping-stones to a deeper and more unified understanding of chemical phenomena.

Similarly, once a critical mass of languages had been described, broader linguistic patterns started to become visible. The linguistic answer to Döbereiner is Joseph Greenberg, who in the 1960s originated the typological approach to the study of language. Greenberg was particularly interested in word order patterns. To study this, he collected a sample of thirty languages from a variety of language families and different parts of the world. Before Greenberg, linguists typically compared a language like Italian with neighboring languages like French, Spanish, and Sardinian, looking for the similarities and differences that related to the languages' shared histories. Greenberg, by contrast, compared Italian to languages like Yoruba and Thai and Guarani, looking for similarities that had nothing to do with any historical relationship. And he discovered some striking patterns.

In Chapter 1 I mentioned that the basic word orders employed by Navajo and Japanese are quite similar, even though the two languages are historically and culturally unrelated. For example, in both languages the direct object comes after the subject and before the verb:

Ashkii	at'ééd	yiyiiłtsą.	(Navajo)
Boy	girl	saw	

'The boy saw the girl.'

John-ga	Mary-o	butta.	(Japanese)
John-SU	Mary-OB	hit	

'John hit Mary.'

Also, in both languages the noun phrase associated with a preposition comes before the preposition (which is therefore properly called a *post*position.)

'éé'   biih   náásdzá.   (Navajo)
clothing into   I-got-back
'I got back into (my) clothes.'

John-ga Mary to   kuruma da Kobe ni itta.   (Japanese)
John-SU Mary with car   by   Kobe to went
'John went to Kobe by car with Mary.'

A third property they share is that a phrase that expresses the possessor of a noun always comes before the possessed noun.

Chidí   bi-jáád   (Navajo)
Car   its-leg
'the wheel of a car.'

John-no   imooto-ga   sinda.   (Japanese)
John-'s   sister-SU   died
'John's sister died.'

English differs from Navajo and Japanese in these respects, but it is far from unique. Its basic word orders are also replicated in historically and culturally unrelated languages, such as the Edo language of Nigeria. In both English and Edo, objects come *after* the verb:

Òzó   mięn   Àdésúwà.
Ozo found   Adesuwa

In both, the noun phrase associated with a preposition comes *after* that preposition.

Òzó rhié   néné èbé   nè   Àdésúwà.
Ozo gave the   book to   Adesuwa

Edo is also the opposite of Navajo and Japanese in that the possessor comes after the possessed noun:

Ọ̀mọ̀   Òzó   rré.
child   Ozo   come
'A child of Ozo's came'; 'Ozo's child came.'

This word order is also a possibility in English (although not the only one).

One might think that these similarities have arisen purely by chance. After all, an object can go in only a limited number of places relative to a verb: either before it or after it. Since there are more than two languages in the world, some will inevitably have the same order of object and verb. What makes Greenberg's results interesting is not that these particular combinations of word orders exist; it is that many other reasonable-looking combinations of word orders do not, or at least are very rare. For example, one can imagine a language that is halfway between English and Navajo in having objects after the verb but also postpositions. A sentence in such a language might look like this:

Chris put the book the table on.

This combination is not found in Greenberg's sample. Conversely, one could expect to find a language with the object before the verb but with prepositions:

Chris the book on the table put.

This is rare in Greenberg's sample. If languages were random collections of specific rules and properties, one would expect such languages not only to exist but to be approximately as common as languages with the Navajo/Japanese or English/Edo word order. But they aren't. Greenberg stated these discoveries in the form of implicational universals, such as the following:

- Universal 3: Languages with dominant Verb-Subject-Object order are always prepositional.

- Universal 4: With overwhelmingly greater than chance frequency, languages with normal Subject-Object-Verb order are postpositional.
- Universal 2: In languages with prepositions, the genitive [i.e., the possessor noun phrase] almost always follows the governing noun [i.e., the possessed noun], while in languages with postpositions it almost always precedes.

Overall, Greenberg found forty-five such "universals" of language, with varying degrees of statistical reliability. This showed clearly for the first time that human languages have similarities that do not emerge from shared culture and history but rather from general properties of human cognition and communication. Contrary to Sapir's view, there *are* assignable limits to the variation found in human languages. Just as Döbereiner's triads pointed to the existence of some deeper underlying structure in the world of chemistry, Greenberg's universals point to the existence of deeper linguistic structure.

Success is to be imitated. Döbereiner's observations led other chemists to try their own organizational schemes on the elements, including Beguyer de Chancourtois's "telluric screw" and John Newlands's music-based system, which revealed an eightfold rhythm to the elements. In the same way, Greenberg's universals have created a cottage industry of typologists looking for similar discoveries. Some have continued to examine approximately the same issues that Greenberg considered, trying to test, refine, and extend his generalizations to a larger sample of languages. For example, Matthew Dryer at SUNY-Buffalo uses a 625-language sample, carefully controlled for history and geography, together with sophisticated statistical techniques, in an effort to distinguish true correlations from accidental similarities.

Others seek universals in areas besides word order. Joanna Nichols, for example, has brought out another important dimension. Whereas Greenberg focused on the word orders of subjects, objects, and verbs, Nichols explores other ways in which languages can dis-

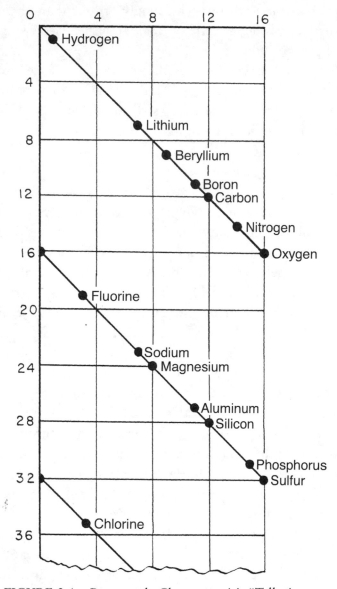

FIGURE 2.1 Beguyer de Chancourtois's "Telluric Screw"—a Precursor to the Periodic Table
SOURCE: *The Discovery of the Elements*, by Willy Ley, © 1968 by Willy Ley. Used by permission of Doubleday, a division of Random House.

tinguish the subject from the object of the same clause. She identifies two. One technique, which we have already seen, is to put case markers on noun phrases that say "this is the subject" or "this is the object." The Australian language Warlpiri involved one example of this. Japanese has another: The subject consistently has the suffix -*ga* added to it, and the object consistently bears the suffix -*o*.

John-ga   Mary-o   butta.   (Japanese)
John-SU   Mary-OB   hit
'John hit Mary.'

If these two suffixes are switched, the meaning of the sentence changes in a predictable way, even with no change in the word order. A Japanese speaker would thus understand *John-o Mary-ga butta* as meaning 'Mary hit John,' not 'John hit Mary.' Other languages, like Mohawk, tolerate flexibility in word order, but for a different reason. In Mohawk no markers are added to the nouns to denote subject and object. Instead, Mohawk changes a prefix on the verb. Compare the following two sentences:

Sak   Uwári   shako-núhwe's.
Jim   Mary   he/her-likes
'Jim likes Mary.'

Sak   Uwári   ruwa-núhwe's.
Jim   Mary   she/him-likes
'Mary likes Jim.'

The prefix *shako-* is used only if the understood subject of the sentence is a masculine singular noun phrase (like *Sak*) and the understood object is feminine singular (like *Uwari*). If the subject is feminine and the object is masculine, as in the second example, then the prefix *ruwa-* is used instead. Mohawk has fifty-eight prefixes of this kind, each of which communicates a different combination of subject and object. Such elements are called *agreement markers* be-

cause the choice of the affix on the verb must agree with the properties of the nouns in the sentence. Linguists often refer to the verb as
the "head" word of a sentence in that it provides the core that the
sentence is built around. Noun phrases are "dependents" of the verb
because the number and kinds of noun phrases to be included depend
on what verb is chosen. With this terminology in mind, Nichols calls
Japanese a *dependent marking* language because it adds affixes to the
noun phrases and Mohawk a *head-marking* language because it adds
affixes to the verb.

So far this is mostly terminological. What makes Nichols's distinction interesting is that languages tend to use head marking and
dependent marking in consistent ways. Consider, for instance, how
Japanese and Mohawk distinguish a possessed noun from its possessor within a noun phrase. In both languages word order is part of the
answer: The possessor usually comes before the possessed noun. But
Japanese also adds a suffix, *-no,* to the possessor noun:

>   John-no   imooto-ga
>   John-'s   sister-SU
>   John's sister

A parallel expression in Mohawk leaves the possessor unmarked but
puts a prefix on the possessed noun that says that its possessor is
masculine and singular:

>   Sak   rao-wise'
>   Jim   his-glass
>   'Jim's glass'

Thus, the way these languages mark possession inside a noun phrase
is parallel to the way they mark subjects and objects inside a sentence.
The possessed noun is the head of the noun phrase because it defines
what is being talked about; the possessor is a dependent that modifies
this head. Given this, Mohawk noun phrases use head marking just as
Mohawk sentences do, and Japanese noun phrases use dependent

marking just as Japanese sentences do. It is unusual for a language to have other combinations of these properties. Languages that mark the noun phrases in a sentence but the possessed noun in a noun phrase are rare at best. Once again, we see that not all imaginable combinations of grammatical properties are permissible. It seems there are deep underlying principles that determine what properties can and cannot occur together in natural languages.

A discipline really gets interesting when the ideas of the thinkers begin to converge with the empirical discoveries of the doers. Such a moment arrived in chemistry with the work of John Dalton. Dalton did not discover any new elements, nor was he the first to propose the concept of the atom. But he was the first to bring the two traditions together in a fruitful way by showing how the long-neglected Greek notion of an atom could explain the discovery that elements combine to make compounds only in exact fixed ratios. For example, starting from the experimental result that water is always made up of 12.6 parts by weight of hydrogen and 87.4 parts of oxygen, together with his hypothesis that individual hydrogen atoms combine with individual oxygen atoms in the simplest way, Dalton concluded that oxygen atoms must be seven times heavier than hydrogen atoms. This convergence between a priori reasoning and exact empirical discovery earns Dalton a special place in the history of chemistry.

Since modern linguistics has a shorter history than chemistry, Chomsky gets to be Dalton as well as Democritus. The two roles were not split, for the linguistic data needed to gave shape and substance to Chomsky's conceptual conclusions came to light without those conclusions' being neglected for centuries. Greenberg's universals or Nichols's typological distinctions could have performed this function. In fact, however, Chomsky's distinctive notion of parameters as the atoms of linguistic diversity emerged out of a specific comparison between French and Italian.

That these two languages should provide the paradigm-forming comparison is partly because of accidents of circumstance and personality. One of Chomsky's first and greatest students, Richard Kayne, began his career by applying Chomsky's early theories to highly detailed analyses of French. He also taught in France for many years. There a young Italian named Luigi Rizzi met him, absorbed this particular way of looking at natural languages, and began the project of translating many of Chomsky and Kayne's crucial examples into his native Italian. There were many similarities—not surprising, given that English, French, and Italian are closely related. But there were differences as well. In a remarkable twist, many of these differences appeared to cluster into a discernible pattern.

The simplest Italian sentences look very much like the simplest French and English sentences: They consist of a subject noun phrase followed by a verb that is marked for tense (an indication of whether the event was past, present, or future).

Jean	arrivera.	(French)
Jean	will-arrive	

Gianni	verrà.	(Italian)
Gianni	will-come	

Italian differs from French and English, however, in that the subject can also come after the verb. Thus, a perfectly acceptable alternative Italian sentence is:

Verrà	Gianni.
Will-come	Gianni

In French, however, one would not say:

*Arrivera	Jean.
Will-arrive	Jean

A second difference comes when the subject of a sentence refers to someone who is already known from the context. In French and English, it is normal in these circumstances not to repeat the name but to use a subject pronoun instead. So if Jean has already been discussed, one would say *Il arrivera* 'He will arrive.' Italian, in contrast, permits a more radical reduction: When the people in a conversation already know whom they are talking about, one can say just the verb with no subject noun phrase at all, not even a pronoun. Thus, *Verrà* qualifies as a complete and well-formed sentence in Italian, meaning 'He or she will come.' **Will come* and **arrivera* do not count as complete sentences in English and French, however.

These first two differences between French and Italian are the kinds of matters one would expect to come up quickly in an Italian 101 class. But the two languages also differ in more subtle ways. English, French, and Italian share a common method of making questions. Some noun phrase in a complete sentence represents the unknown information. That noun phrase is replaced by a suitable question word, and the question word is moved to the front of the sentence. For example, suppose that we wanted to make a question corresponding to the following piece of information:

Chris will see someone in the park.

One substitutes a question word for the unknown part:

Chris will see whom in the park?

Then one relocates the question word at the beginning of the sentence, leaving a gap where the object would normally be:

Whom will Chris see _____ in the park?

(Note that the future tense auxiliary *will* also shifts to come before the subject in this sentence. This detail is peculiar to English, and I ignore it here.) But this simple procedure has limitations in English and French. Consider a more complex initial sentence, in which one clause appears embedded inside another:

You said that Chris saw Pat in the park.

Tu	veux	que	Marie	épouse	Jean.	(French)
You	want	that	Marie	marry	Jean	

If one replaces the object of the embedded clause with a question word, the rule works as before in both languages:

Whom did you say that Chris saw _____ in the park?

Qui	veux-tu	que	Marie	épouse	_____?	(French)
Whom	want-you	that	Marie	marries?		

If one replaces the *subject* of the embedded clause with a question word and moves it to the front with no extra changes, however, the result is ungrammatical:

*Who did you say that _____ saw Chris in the park?

*Qui	veux-tu	que	_____	épouse	Jean?	(French)
Who	want-you	that	_____	marries Jean?		

These sentences need to be fixed by changing the word that introduces the embedded sentence: In standard English the conjunction *that* must be omitted; in French the parallel word *que* must be changed to *qui*.

Who did you say _____ saw Chris in the park?

Qui	veux-tu	qui	_____	épouse	Jean?	(French)
Who	want-you	that	_____	marries Jean?		

In Italian, in contrast, the normal rule that question words move to the beginning applies in its most pristine form even to subjects of embedded clauses. Italian also has a designated word for introducing embedded clauses, namely, *che* (cognate to the French *que*):

Credi	che	Gianni	verrà.
You-think	that	Gianni	will-come

Che is neither deleted nor changed when one questions the embedded subject in Italian:

Chi	credi	che	_____	verrà?
Who	you-think	that	_____	will-come?

This is a much subtler but still significant difference between Italian grammar and the grammars of French and English. One can imagine even completely fluent speakers making mistakes on this point, although native speakers would notice this as an error.

Why did Kayne and Rizzi argue that all these differences between Italian and French (and English) are interrelated? Why aren't they just three random differences, each independent from the others? There are at least three reasons. The first involves comparison with other Romance languages, such as Spanish and Romanian. Spanish also allows the subject to follow the verb, allows "redundant" subjects to be omitted, and allows the subject of an embedded clause to be questioned with no readjustments. In all these respects, Spanish is grammatically more like Italian than French, even though by the normal criteria of historical linguistics it is more closely related to French. At the other edge of the ancient Roman Empire is Romanian. Although it has been isolated geographically from the other Romance languages for centuries, it also behaves systematically like Italian rather than French in these respects. If these three properties were unrelated, one would expect them to appear in more or less random mixtures in the various Romance languages. But that is not what we find.

A second reason to think that these facts are interrelated comes from historical linguistics. We have extensive written records in French that can be used to trace its history in considerable detail. Examination of older texts reveals that earlier stages of French were like Italian and Spanish for all three of these properties and that each of these Italian-like features (originally derived from Latin, the common ancestor) phased out between Middle French and Modern French. Thus, French changed from being a pure Italian-type language to

e French-type language over a century or two. That these
changed together and at roughly the same time gives fur-
nce to the claim that they are all related.

These first two reasons would not by themselves have led Chomsky
and his colleagues to the idea of a parameter, however. The third and
most important reason for thinking these differences are interrelated is
that all three have a discernible common theme: All involve the subject
in one way or another. That does not seem like a coincidence.

Indeed, we can be more precise. English and French have a requirement that (almost) every clause with a tensed verb must have a
visible subject of some kind. The most striking demonstration of this
comes from sentences that concern the weather. These typically begin
with *it* in English or the equivalent *il* in French:

Il pleut.
It is raining.

Superficially, these sentences look just like *It is boiling,* but there
is an important difference. The *it* in *It is boiling* is a kind of abbre-
viation for a full noun phrase that can be recovered from the context.
It is boiling means *The soup on the stove is boiling* in a situation
where we needn't bother saying *the soup on the stove.* The *it* of *It is
raining* is not short for anything. This sentence can be used without
any physical gesture or earlier sentence to define what *it* is. Indeed,
sentences in which an ordinary noun phrase appears as the subject of
these verbs sound very strange:

*The cloud is raining.
*The weather is raining.
*Montreal is raining.

Apart from metaphorical uses such as *Confetti rained down on the he-
roes,* we do not think of raining as the sort of activity that things do;
it is something that just happens spontaneously. There is usually no semantic subject for a verb like *rain.* Nevertheless, English and French
have such a strong requirement that tensed clauses have subjects that

speakers feel a need to make one up. Since *it* and *il* are the noun phrases in these languages with the least inherent meaning, they are pressed into duty. This shows that the need for a subject is a grammatical requirement that holds even when there is no semantic subject to talk about. Italian and Spanish are different in this respect; statements about the weather in these languages show up as bare verbs.

Piove.   (Italian)
Llueve.   (Spanish)
Is-raining

Italian and Spanish speakers do not feel the same compulsion to make up a subject just to have one. We thus have a fourth basic difference between Italian-like languages and French-like languages, which can be stated as follows:

In some languages (French, English, the Nigerian language Edo, etc.) every tensed clause must have an overt subject.
In some languages (Italian, Spanish, Romanian, Japanese, Navajo, etc.) tensed clauses need not have an overt subject.

What is important about this is that the other three differences between French and Italian I enumerated above can all be seen as consequences of this one fundamental difference. The most obvious application is that one must say *She will come* in English and French whereas a simple *Verrà* will do in Italian. This example is slightly different from the weather sentences, because here the subject pronoun is not a mere placeholder but stands for a real, meaningful noun phrase. Here, too, however, Italian and Spanish speakers need not bother with a subject, whereas French and English speakers must include at least a pronoun.

Consider next that subjects may come after the verb in Italian: *Verrà Gianni* ('comes Gianni') is possible as well as *Gianni verrà*. I implied above that such reversals of word order are not possible in English and French, but that was not quite accurate. A similar change is sometimes possible in these languages—but a dummy pronoun must appear in the normal, preverbal subject position. Although one cannot say *Appeared a boat* in English, then, or *Est arrivé Jean* in French, one can say:

There appeared a boat on the horizon.

Il	est	arrivé	trois	hommes.	(French)
It	is(has)	arrived	three	men	

The difference between Italian and both French and English is not exactly what we thought at first. The "logical subject" can be bumped out of its usual position in all three languages, but in French and English a new subject must be added, to satisfy the condition that any clause that has a tensed verb also has a subject. This requirement is not active in Italian. Therefore, no dummy subject pronoun appears in inverted sentences in Italian, just as no subject pronoun is needed with weather verbs or verbs whose subject is recoverable from context.

The last contrast we observed had to do with questioning the subject of an embedded clause. In Italian this is possible with no adjustments, whereas in French and English it is not:

Chi	credi	che	_____	verrà?	(Italian)
Whom	you-think	that		will-come?	
*Whom did you say that _____ will come?					
*Qui	veux-tu	que	_____	vienne?	(French)
Whom	want-you	that	_____	come?	

Now we can see why this difference arises. Moving a question word to the front of the sentence again leaves behind a tensed clause with no overt subject—the configuration that we know is easily tolerated in Italian but not in French or English.

This proposal elegantly explains why questioning an embedded subject is problematic in French and English, but other questions are not. For example, the difference doesn't show up when one questions the subject of a simple sentence like *Who will come?* Such questions are equally possible in all three languages. Here the question word didn't need to move anywhere to get to the front of the sentence. It can stay in the normal subject position, satisfying the French and English requirement. We can also explain why no difference shows up when one questions the *object* of an embedded clause. Crucially, there is no requirement that clauses in French and English have direct objects; English speakers feel no need to add a dummy object to *rains* to form *It rains it,* for example. Question movement can thus remove the object from an embedded clause freely, with no adjustments in any of these languages. (For instance, *Chris thinks that I bought a dog* easily becomes *What does Chris think that I bought?*) Thus, even though the grammatical rule about subjects says nothing about questions directly, it has implications for questions that are predictably different in the different languages.

This comparison between English and French versus Italian and Spanish is what led Chomsky to propose the idea of a parameter. At first glance these languages seem to differ in many ways. Some of the differences are obvious to anyone trying to learn these languages; others are extremely subtle and had never been noticed until linguists realized what to look for (including some differences I have not presented here). Chomsky realized, however, that French and Italian actually diverge only in a single feature, which expresses itself differently in different grammatical constructions. If there were really six differences instead of one (the four I discussed, plus two others I omitted), we would expect them to vary independently from each other, resulting in (approximately) $2 \times 2 \times 2 \times 2 \times 2 \times 2 = 64$ different kinds of Romance languages. In reality there are only (approximately) two kinds of Romance languages in these respects: the French type and the Italian/Spanish/Romanian type. If there were really six differences instead of one, it would be surprising that French changed from a uniformly Italian-type language to a uniformly English-type language. Moreover, the single feature can be isolated and identified by linguistic analysis: The difference, whatever it is exactly, has to do with whether sentences need subjects or not. Chomsky called such a single feature a *parameter*. More specifically, this particular feature is referred to as the *null subject parameter*, for obvious reasons. Such parameters can be "set" in at least two ways: For example, a language might require subjects, or it might not. The collection of observable consequences that follows when a language chooses one setting of a parameter rather than another is known as a *parametric cluster*.

Chomsky further observed that parameters like this could take away some of the mystery of how children can learn something as complex as a natural language so easily and reliably. One cannot expect all English-speaking children to notice that adults don't say sentences like *Who does Pat think that will marry Chris?* whereas they do say sentences like *Whom does Pat think that Chris will marry?* Children have relatively little exposure to sentences this complex, particularly in the early stages of learning a language. One can, however, expect children to notice whether one uses a subject pronoun in a sentence like *It is raining*. Once children learn this, they can deduce that subjects must be obligatory in English. Similarly, when children in Italy hear sentences like *Piove*, they can conclude that subjects are not obligatory in Italian. Without needing any direct exposure to the exact structures, both groups of children are then able to infer (unconsciously) the consequences of this rule for complex sentences involving embedded clauses and questions. Because the differences among languages cluster into stable patterns, children can learn parts of the pattern indirectly, as a consequence of learning another, more accessible part. Parameters thus simplify the logical problem of language acquisition enormously. Encouraged by this, Chomsky made the bold conjecture that this difference between French and Italian is typical. In 1981 he proposed that all differences among languages are to be thought of in this way, as different choices that languages make with respect to a finite number of parameters.

———————

The atomic theory in chemistry made the bold and surprising claim that the vast diversity of substances we observe can be characterized as different arrangements of a smallish number of discrete elements. The parametric hypothesis makes a similar claim: The diverse array of languages we observe can all be characterized as different arrangements of a smallish number of discrete parameters. Both hypotheses treat what looks like a continuous analog-style phenomenon as being essentially digital. Furthermore, the original Greek word for 'atom' means 'uncuttable': It implies there is a smallest unit of matter that cannot be further subdivided. Similarly, parameters create parametric clusters that are also in a sense uncuttable. The null subject parameter of Italian is an irreducible feature of that language; it should not be cut apart into smaller features such as an ability to omit pronouns and an ability to question embedded subjects. Just as atoms gave chemistry a way of resolving its foundational paradoxes, so parameters give linguistics a way of resolving its foundational paradoxes of similarity and difference. Parameters are the atoms of linguistic diversity.

Some people, particularly some language scholars, have been scandalized by these proposals. Chomsky's own research has focused mostly on the details of English grammar, together with some comparison with closely related languages such as French, Italian, and Spanish. How, these critics ask, can one infer anything about a universal grammar or the nature of all human languages from such a limited sample? The enterprise seems to smack of ethnocentrism, of the presupposition that everything must be like English. It also seems not to do justice to the continuous variation that the theory's opponents think they see in language. Still, we know from watching a digitally mastered movie or trying to catch a glimpse of the droplets in a stream of water coming from a faucet that things that are made up of discrete pieces can easily look continuous to our unaided senses. Moreover, the "limited sample" argument is specious: When one needs a new idea for addressing a paradoxical situation, more facts often just add to the confusion. What is really needed is a new perspective on the observations already in hand. Chomsky's conclusions about universal grammar and the existence of parameters are no more astonishing than Democritus' conclusions about the existence of atoms. Democritus, too, observed only a tiny number of the substances known to modern chemistry. By the standards of the scientific method and modern instrumentation, he didn't even observe them very closely. But he did think about what he saw and reached a correct conclusion by valid reasoning.

I should confess that not every linguist would assign so much importance to the notion of a parameter or understand it in exactly the same way. Within the Chomskyan paradigm, there are many linguists who accept the terminology of parameters but have somewhat different views about exactly what a parameter is and what the best examples are. Outside the paradigm, many linguists object (sometimes strenuously) to the terminology of parameters and some of the intellectual background associated with it, preferring a different terminology and different associations. But beneath the surface of controversy and debate, there is a growing understanding that the differences among languages are to be grouped into relatively stable patterns that do not arise as accidents of particular histories or cultures.

Where does this leave the historical development of linguistics? It is always tricky to understand the present. But in light of the growing awareness that something like the parameter exists, linguistics in my view is ready and waiting for its Mendeleyev.

Chemistry had to get ready for Mendeleyev for some time. One cannot hope to have a correct and explanatory classification of the chemical elements until one has discovered a certain percentage of those elements and observed their basic properties. Chemists needed to calculate various atomic weights and discover similarities in various chemical reactions. The basic concept of an atom had to be in place, to provide an effective way of thinking about both atomic weights and the formation of chemical compounds. These basic preconditions were satisfied in the middle of the nineteenth century. Since the 1820s, scientists like Döbereiner had begun to discern that sometimes there were systematic relationships between an element's atomic weight and its chemical properties. By 1860 an important confusion over how to calculate atomic weights had been resolved. As a result, chemists called a congress to be held in that year at Karlsruhe, Germany, to explore the possibility of using real atomic weights as a comprehensive classification system for the chemical elements—a system that could bring order to their ever-increasing knowledge about those elements. This desire was stimulated by their envy of the zoologists and the botanists, who already had classification schemes that encompassed all living things.

Dmitry Mendeleyev, who attended the Karlsruhe congress, deliberately took up this project. When he returned to his native Russia in 1861, one of his responsibilities was to write a textbook. He wrestled with the question of how to present his knowledge in an orderly and systematic way that would be easy for his students to understand. To this end, he spent eight years gathering information, writing countless letters to research centers all over Europe to get the best figures on atomic weights and other numerical properties. He also used his love of solitaire games by writing the name of each element on a playing card and dealing the cards out in endless arrangements. Finally, in 1869, he hit upon the scheme of ordering the elements by their atomic weights into two short "periods" of seven elements each, followed by three long "periods" of seventeen elements, with some gaps left for elements that had not been discovered yet. His breakthrough was largely ignored at first, until the better-connected German Lothar Meyer published the beginnings of the same system in 1870. The primacy and greater completeness of Mendeleyev's work was soon recognized, however. When in the ensuing years three of Mendeleyev's missing elements were discovered and found to have properties that accurately matched Mendeleyev's predictions, chemists began fully to appreciate the genius of his system. Some inadequacies needed to be fixed: Mendeleyev did not know where to fit in hydrogen, he was completely ignorant of the noble gases, and he made wrong predictions as well as right ones when it came to sorting out the rare earth metals. But the basic vision was sound. Today nearly every chemistry textbook, classroom, and laboratory has a version of Mendeleyev's periodic table displayed prominently. The periodic table is special because every natural element is included, each in its proper place in relation to all the others. All the

TABLE 2.1   Mendeleyev's Periodic Table

Groups	Higher Salt Forming Oxides	Typical 1st Small Period	Large Periods				
			1st	2nd	3rd	4th	5th
I.	R_2O	Li = 7	K 39	Rb 85	Cs 133	—	—
II.	RO	Be = 9	Ca 40	S 87	Ba 137	—	—
III.	R_2O_3	B = 11	Sc 44	Y 89	La 138	Yb 173	Th 232
IV.	RO_2	C = 12	Ti 48	Zr 90	Ce 140	—	—
V.	R_2O_5	N = 14	V 51	Nb 94	—	Ta 182	Ur 240
VI.	RO_3	O = 16	Cr 52	Mo 96	—	W 184	—
VII.	R_2O_7	F = 19	Mn 55	—	—	—	—
VIII.			Fe 56	Ru 103	—	Os 191	—
			Co 58.5	Rh 104	—	Ir 193	—
			Ni 59	Pd 106	—	Pt 196	—
I.	R_2O	H = 1. Na = 23	Cu 63	Ag 108	—	Au 198	—
II.	RO	Mg = 24	Zn 65	Cd 112	—	Hg 200	—
III.	R_2O_3	Al = 27	Ga 70	In 113	—	Tl 104	—
IV.	RO_2	Si = 28	Ge 72	Sn 118	—	Pb 206	—
V.	R_2O_5	P = 31	As 75	Sb 120	—	Bi 208	—
VI.	RO_3	S = 32	Se 79	Te 125	—	—	—
VII.	R_2O_7	Cl = 35.5	Br 80	I 127	—	—	—
		2nd Small Period	1st	2nd	3rd	4th	5th
					Large Periods		

basic properties of those elements are expressed. Indeed, the most important properties are expressed in a particularly natural way, as Mendeleyev's table revealed for the first time the correct relationship between atomic weight and chemical valency. In this respect, the periodic table surpassed the zoological and botanical classification systems, which express some important biological relationships but not others. Mendeleyev's ability to recognize unknown elements as gaps in the table and make precise predictions about them was also unprecedented. No zoologist could predict the existence of a new animal based on his observations of known animals, but Mendeleyev was able to predict that there would be such a thing as germanium from his analysis of silicon. These striking successes were possible because chemistry had, for the first time, a theory of what combinations of properties an element could have and of what logically possible combinations an element could never have. There cannot, for example, be a halogen that forms compounds similar to those of chlorine but that has an atomic weight between those of aluminum and silicon. By extension, the table also expresses what kinds of compounds can in principle be built from elements, given their valence. This theory brought order to what was known, revealed its underlying symmetries, and made strikingly correct predictions about not-yet discovered elements. It remains the organizing principle of chemistry today, standing as a landmark in the maturing of chemistry as a science and as one of the great achievements of the human mind.

Linguistics now seems to be in a stage similar in many ways to where chemistry was in the mid nineteenth century, just prior to Mendeleyev. The key theoretical idea of the parameter is in place, together with an appreciation of how it can be used to solve linguistic problems. We also have practical experience with a certain number of actual parameters. The null subject parameter seems to be one. At least one more is lurking in Greenberg's universals concerning word order: Languages seem to make one choice that determines whether verbs will come before their objects, prepositions before their noun phrases, and nouns before their possessors, or whether it will be the other way around. A third parameter is to be found in Nichols's distinction between head-marking languages and dependent-marking languages: Languages choose whether verbs will bear agreement affixes that are determined by the noun phrases or whether noun phrases will have case affixes that are determined by the verb. There are others as well. We even have some understanding of how these different parameters can interact with each other to give more complex patterns, comparable to chemists' knowledge of how atoms can combine to form compounds. Thus, we are approaching the stage where we can imagine producing the complete list of linguistic parameters, just as Mendeleyev produced the (virtually) complete list of natural chemical elements.

This list of parameters will be sufficient to characterize the grammatical skeleton (though, of course, not the "skin" of pronunciation, idiom, or figure of speech) of any natural human language, ancient or contemporary. Properly organized, such a list will constitute a kind of periodic table of languages. We might even be in a position to describe possible languages that no one has yet observed. This will happen when we recognize that the known parameters can combine in some logically consistent way to form a language with distinctive properties that is so far unknown. Then we can hope that this prediction will be confirmed by some linguist working with a yet undescribed language from a remote jungle or a philologist considering an ancient text or a dialectologist looking at a newly formed speech variety. But even apart from the predictive power it might have, this imagined periodic table of languages will have value in succinctly and gracefully summarizing what is known about languages, including what is possible and what is not, in a way that reveals the true elegance under the bewildering wealth of facts. Then one great historical thrust of linguistics will have reached fulfillment, bringing the field to a new level of maturity. With a coherent organization of the atoms of language and their modes of combination, we will be ready to move on—to discover the linguistic equivalents of radioactivity and quantum mechanics, whatever those turn out to be.

3

Samples Versus Recipes

WHEN MOST PEOPLE HEAR THE PHRASE *atoms of language,* what comes to their minds is words. Words are the little pieces of language that we consciously look up in dictionaries and piece together into sentences when we write. Learning vocabulary is perhaps the largest and most laborious aspect of acquiring another language. Basic words are like the Greek notion of atoms in that they cannot be divided into smaller meaningful parts. Therefore, it seems that words must be the atoms of language. Where, then, does talk of parameters come in?

This common reaction is correct—in one sense. But like most terms we use in everyday speech, *language* has more than one sense. You can get a flavor for this by opening any largish dictionary to a random page and scanning the entries. Almost every word listed has multiple meanings, and even so dictionaries do not capture every technical sense or extended use that is adapted to the needs of the moment. For example, when physicists or engineers speak of *work,* they are not talking about whatever you do to get a paycheck but about force exerted in the direction of motion times distance. These two senses of work are not entirely unrelated (there are reasons physicists picked this label for a key concept in their theories), but they are certainly not the same. Similarly, in this chapter I contrast two rather different meanings of the word *language* that are important for linguistics, arguing that the Code Talker paradox arises largely from a failure to distinguish these two senses. Although words are indeed the atoms of language from one perspective, parameters are the atoms of language from the second perspective, playing the same explanatory role in linguistics that atoms play in chemistry. Sharpening our understanding of what a language is will enable us as well to be more explicit about exactly what parameters are. I also present a case study of a particular parameter in detail, to show how it dissolves one part of the Code Talker paradox.

The primary distinction that needs to be made can be approached by an analogy. My family's favorite version of chemistry is the kind that happens in the kitchen. Although I inherited the name *Baker* from my ancestors, visitors to our home soon realize that my wife, Linda, is the real baker. Along with certain talents at desserts that do not concern us here, she makes virtually all of our family's bread, according to a procedure she has developed and refined over many years. Our dinner guests often make the mistake of not paying much attention to the bread at first. At some point during the meal, however, it usually dawns on them that this bread is special. Sometimes their enjoyment of it even distracts them from the other dishes. At the end of the meal, they often wish they could have more bread later.

There are two quite different ways that we can (and do) respond to their desire. On the one hand, Linda can offer them a *sample* of the bread: an extra loaf that they can take home with them and eat the next day. On the other hand, she can offer them her *recipe:* the list of ingredients and the proper procedures for combining them in order to make the bread. In both cases we can say that Linda is giving her bread to our guests. But she is giving the bread in two quite different senses, a difference reflected in the ways people respond to the two offers. To those endowed with the right kind of bravery and initiative, the recipe is the more attractive offer, because it opens up the possibility of an unlimited supply of the bread once they learn to make it for themselves. Most people, however, gratefully accept the sample loaf and leave it at that. They can enjoy it the next day without exercising any particular bravery or initiative outside their usual routine. The loaf will not last long, but in the short term it tastes much better than the index card.

This distinction between samples and recipes can be applied to many things. Consider a mathematical example. Suppose that I wanted to communicate to you a sense of what even numbers are. I could do this by giving you samples of even numbers: The even numbers are 2, 4, 6, 8, 10, and so on. In set notation, the even numbers can be expressed as {2, 4, 6, 8, 10, . . . }. In the jargon of logic, this is called an *extensional* characterization of the set of even numbers. Alternatively, I could give you a recipe for making even numbers: An even number is any number that is the result of multiplying an integer by two. Set notation expresses this as $\{x: x = 2y, y$ an integer$\}$. Here I have not given you an example of any even number, but I have given you a way of identifying all of them. This is called an *intensional* characterization of the set of even numbers; it points to a property that all even numbers share by virtue of being members of the set. The two sets are identical, even though the descriptions are different.

Human languages can be thought of in these two ways as well. Suppose someone were to ask you what English is. You might respond by pointing to examples of written or spoken English. For example, you could hand her this book, open to this page, point to the ink marks and say, "This is English." This would be an extensional characterization of English: You are calling attention to some representative samples of actual English sentences. The other choice is to give your questioner some kind of recipe for forming and recognizing English sentences from scratch. You might say that English is the set of sentences that are constructed by combining the following ingredients (you hand her a massive dictionary that lists all the English words) according to the following rules of grammar (you hand her an equally massive English grammar). If these manuals were written in (say) Japanese, they might contain no complete English sentences at all; yet between them they would tell the reader how to make any conceivable English sentence. This would be an intensional characterization of English because it is a description of the properties that all English sentences have.

Giving a complete intensional definition of English would be impractical, in part because no complete grammar of English exists. But the extensional definition is no better, because any sample of English you might point to would also be highly incomplete. Though we lack complete word lists or grammatical rules for English, at least we can conceive of them, and we have things that approach that ideal. But it would be impossible, even in principle, to produce a finite list of the sentences of English, since new sentences are always being created.

When we speak informally of *English,* we could have either of these two senses in mind. We could be thinking of English-as-recipe, or we could be thinking of examples-of-English. In one sense, English is a procedure for making sentences. This is what we implicitly mean when we say of someone that he *knows* English. We cannot mean by this that he has in mind a list of all possible English sentences. We must mean instead that he can follow the recipe for producing (and understanding) English sentences. Chomsky, who has done much to clarify this distinction, calls this sense of language "I-language," where *I* reflects the fact that it is an *intensional* characterization of language, one that is *internal* to the mind of a speaker. Alternatively, we could be thinking of English as a collection of actual sentences. That might be what we implicitly have in mind when we say of people that they *speak* English. By this we mean that many of the sentences that come out of their mouths are examples of the English language. Chomsky calls this sense of language "E-language," where *E* reflects the fact that it is an *extensional* notion of language, looking at *examples* of language as they exist *external* to the minds of people who speak the language.

It is not always clear or especially important which of these two senses of English we have in mind. Even when words are technically ambiguous, we need not always resolve the ambiguity fully. I do not intend my readers to feel tortured because whenever they hear someone use the word *English* they feel compelled to wonder, "Does that person mean English as an I-language or English as an E-language?" Still, understanding this distinction is crucial to making progress on the paradoxes of language diversity laid out in Chapter 1. There we asked whether two historically unrelated languages, such as English and Navajo, are mostly the same or mostly different. Some facts seemed to point toward one answer, whereas others pointed to its opposite. Now that we have recognized an ambiguity in what we mean by language, this paradox becomes easier to understand. The first step is to clarify the question: Are we interested in comparing E-English to E-Navajo, or are we interested in comparing I-English to I-Navajo? Are we comparing English and Navajo extensionally, by looking at actual sentences, fully formed? Or are we comparing them intensionally, asking whether the procedures for making English sentences and the procedures for making Navajo sentences are similar or not? Are we comparing the recipes for the two languages or the products of those recipes?

The two versions of these questions may have two different answers. Intensional definitions can be quite similar even when the extensions are very different. As a simple mathematical example, compare the even numbers with the set of multiples of seven. From the perspective of extensional lists, the two sets don't much resemble each other: 7, 14, 21, 28 . . . doesn't look like 2, 4, 6, 8 . . . . Only a rather small percentage of the numbers on the list of even numbers is also on the list of multiples of seven (one-seventh of them, to be exact). Yet from the perspective of an intensional recipe, the two sets are extremely similar: They are constructed in exactly the same way, except that you multiply by a different factor. In set notation the even numbers are $\{x: x = 2y, y$ an integer$\}$, and the multiples of seven are $\{x: x = 7y, y$ an integer$\}$. The two recipes are identical except for one symbol. This accurately captures the fact that the two sets have many mathematical properties in common. Indeed, the constant factor in these equations is a *parameter* in the original, mathematical sense of the term. It is this usage that gave Chomsky the idea of using the word in linguistics as well.

In other domains the difference between comparing samples versus comparing recipes can be even more dramatic. This will happen whenever there is a significant interaction between the ingredients in a recipe that is not directly proportional to the quantities of those ingredients. For example, to the uninitiated, a cracker and a loaf of bread seem like quite different foods. Yet the recipe for bread is very much like the recipe for crackers: One merely adds a tablespoon of yeast to get bread. This small difference in the ingredient list makes a striking difference in the qualities of the final product, since the yeast has a dramatic interaction with the other bread ingredients. As St. Paul observed long ago when reflecting on the Jewish Passover, "a little leaven leavens the whole lump." Similarly, a dash of chili powder or a single jalapeño pepper can go a long way in transforming the taste sensations produced by many kinds of food.

Something analogous happens in the domain of language. The recipes for speaking English and Navajo are surprisingly similar, once they are properly understood. But a few small changes in the recipe interact in complex ways with the other ingredients of the language. The result is that any sample of English looks quite different from a corresponding sample of Navajo. I-English is very much like I-Navajo, but E-English seems very different from E-Navajo. As St. Paul might have said to a different audience, a little parameter parametrizes the whole language.

This idea can resolve the Code Talker paradox. If the evidence we have for the similarity of languages can be understood as tapping into the concept of I-language, and the evidence for extreme differences among languages taps into the concept of E-language, the paradox is resolved. This seems to be the case. Consider, for example, the surprising fact that well-educated and well-funded cryptographers could not decode the Navajo language, whereas a young missionary child could. Presumably the cryptographers were attempting to find various mathematical transformations that they could apply to the Navajo sentences to turn them into English (or Japanese) sentences. If so, they were thinking in terms of E-language, looking for sentence-by-sentence correspondences without making explicit reference to the minds of the speakers. In those terms, the problem is extremely complex. In contrast, children learning a language they hear around them are clearly attuned to the problem of I-language. They are trying to discern the recipe for Navajo sentences so they will be able to produce and understand new ones at will. That problem can apparently be solved with relative ease if one knows how to do it.

Because it can reveal the hidden similarities among languages, the notion of I-language is the central topic of much contemporary linguistic research. Once this is understood, it should be relatively easy to see what a parameter is: A parameter is simply a choice point in the general recipe for a human language. A parameter is an ingredient that can be added in order to make one kind of language or left out in order to make another kind. A parameter could also be a combining procedure that can be done in two or three different ways to give two or three different kinds of languages. If you take the generic ingredients of language, add spice B, and shake, you get English. If you take the same basic ingredients of language, but instead of spice B you add flavorings D and E and stir, you get Navajo. I-languages are recipes, and parameters are the few basic steps in those recipes where differences among languages can be created.

This discussion should clear up the potential confusion over whether the atoms of language are words or parameters. Words are the atoms of E-language—the sense of *language* that people often think of first. When we consider an actual paragraph of English, the smallest bits that the paragraph can be divided into that are still recognizably English are (roughly) the words. That is why dictionaries are so prominent in the reference sections of libraries. But this process of analysis crucially starts with a sample of English, seen as a set of sentences characterized apart from the mind that formed them. We must (also) think of languages as I-languages, as recipes for making sentences. And the smallest parts of a recipe or an algorithm are the individual instructions that constitute the recipe. Thus, words may be the atoms of E-language, but parameters are the atoms of I-language.

So far we have seen exactly one actual parameter: the null subject parameter. It might be stated informally as follows:

In some languages every tensed clause must have an overt subject
    noun phrase.
In other languages tensed clauses need not have an overt subject
    noun phrase.

Now we can recognize what kind of thing this parameter is and in
what view of language it is embedded. It is clearly a rule, an instruction for building a clause. It specifies whether or not a particular
grammatical ingredient—a subject—is needed to form another kind
of linguistic entity (a clause or sentence). As such, it belongs to an I-
language-style characterization of a language, not an E-language
characterization. This parameter is, as advertised, an elementary part
of the recipe for particular languages.

    Let us now look in some detail at another of the basic parameters
of language: the parameter that determines the basic order of words
in a language. In addition to giving us a second instance of a parameter, this extended example provides a striking illustration of how
languages that look very different in terms of E-language can nevertheless be very similar in terms of I-language.

    In some languages it doesn't matter what order the words are spo-
ken in, but in many it does. It is a remarkable discovery of Joseph
Greenberg and other early typologists that most languages of the latter type use one of two basic word orders. One of these is illustrated
rather nicely by the following English sentence:

> The child might think that she will show Mary's picture of John to
>   Chris.

The alternative order is illustrated by the following Japanese
sentence:

> Taroo-ga   Hiro-ga Hanako-ni   zibun-no   syasin-o   miseta to
>   omette iru.
> Taro-SU   Hiro-SU   Hanako to   self-POSS   picture-OB showed that
>   thinking be
> 'Taro thinks [literally, is thinking] that Hiro showed a picture of him-
> self to Hanako.'

The differences between the Japanese arrangement of words and the English one are so severe that if English speakers hear the word-by-word translation of the Japanese sentence, they will find it hard to grasp what the sentence is supposed to mean. It sounds like gibberish. A pattern emerges, however, if one isolates particular differences and considers them one by one.

First, these two sentences illustrate again three characteristic differences between English and Japanese that I discussed in Chapter 2. For one, in the English sentence the direct object noun phrase *Mary's picture of John* comes immediately after the main verb of the embedded clause, *show,* whereas in the Japanese sentence the direct object, *syansin-o* 'picture,' comes immediately before the embedded verb *miseta* 'show.' Second, in the English sentence the noun phrase *Chris* comes after the preposition *to* that it is the object of, whereas in the Japanese sentence the corresponding noun phrase *Hanako* comes before the associated postposition *ni* 'to.' (A noun phrase can consist of only a single word in some cases; I return to this later in the chapter.) Third, in the English sentence the noun *picture* comes before the prepositional phrase *of John* that expresses what is portrayed in the picture, whereas in the Japanese sentence the parallel element *syasin* 'picture' comes after the postpositional phrase *zibun no* 'self of.' Three important word order contrasts are thus illustrated in these two complex sentences.

Further consideration of these sample sentences shows that the tendency of English and Japanese to use opposite word orders goes even deeper. The English sentence contains a modal/tense auxiliary *might* that comes before the main verb, *think.* Similarly, in the embedded clause the auxiliary *will* comes *before* the main verb, *show.* In contrast, the auxiliary *iru* 'is' in the Japanese sentence comes after the main verb, *omotte* 'thinking.' In the English sentence, the embedded clause *she will show Mary's picture of John to Chris* is *preceded* by a special clause-introducing conjunction, *that,* which linguists call a *complementizer.* This clause as a whole then *follows* the main verb, *think,* to which it is related. In the Japanese sentence, the embedded clause is *followed* by a special clause-concluding complementizer *to* 'that,' and the embedded clause as a whole *precedes*

TABLE 3.1   Word Order Relationships in English and Japanese

Element A	Element B	English Relation	Japanese Relation
Verb	Direct object	A precedes B	A follows B
Verb	Pre/postposition phrase	A precedes B	A follows B
Verb	Embedded clause	A precedes B	A follows B
Pre/post-position	Related noun phrase	A precedes B	A follows B
Noun	Related pre/post phrase	A precedes B	A follows B
Complementizer	Embedded clause	A precedes B	A follows B
Auxiliary	Main verb	A precedes B	A follows B

the main verb, *omotte* 'think.' Finally, in the English sentence the prepositional phrase *to Chris follows* the verb *show* that it is semantically related to. In contrast, in the Japanese sentence the postpositional phrase *Hanako ni* 'to Hanako' *precedes* the verb that it is semantically related to, *miseta* 'show.' In all, there are at least seven distinguishable contrasts between English word order and Japanese word order, as summarized in Table 3.1. These statements are true not only of these particular sentences in English and Japanese but hold consistently for each language as a whole. Every verb in English comes before its direct object (if it has one), every postposition in Japanese comes after the related noun phrase, and so on.

These word order patterns are valid not only for English and Japanese but for the majority of the other languages in the world. Indeed, there are languages with English-like and Japanese-like word orders (occasionally with minor variations) on every continent. The following sentence from the Nigerian language Edo shows a word order that is essentially the same as English, even though the people had no contact with English speakers until modern times. Because of this syntactic similarity, an English speaker can read the word-by-word translation of the Edo sentence and understand it immediately.

Òzó má   tá wéẹ́ írẹ̀n ghá rhiè éfótò Úyì yè néné ékpétìn.
Ozo did-not say that he will put photo Uyi in the box
'Ozo did not say that he will put a photo of Uyi in the box.'

All seven of the English word order generalizations are illustrated in this Edo sentence, as interested readers can check. Other languages with similar word order characteristics include Austro-Asiatic languages such as Thai and Khmer, some of the Austronesian languages spoken in the Pacific Islands (including Indonesian), many of the Chinese languages (although with some complications), most of the Niger-Congo languages of sub-Saharan Africa, and some aboriginal languages of the Americas, including the Zapotec languages of Mexico and the Salish languages of the Pacific Northwest. Thus, there is nothing uniquely English about "English-style" word order. Neither is the alternative word order to be seen as a peculiarity of Japanese culture, caused by eating too much raw fish or by years of devotion to the martial arts. It is also found in the languages of people with diverse cultures from around the world. Other languages with Japanese-style word order include the Turkic languages, the Dravidian languages of South Asia, many languages of New Guinea, some languages of the Caucus Mountains, some African languages such as Amharic (spoken in Ethiopia), the Basque language (Europe's one surviving pre-Indo-European language), many languages of the American Southeast and Southwest (including Navajo), the languages of the Eskimos, and Quechua, the language of the ancient Incan empire in South America. The following sentence shows Japanese-style word order in Lakhota, the language of the Sioux Indians:

John wowapi k'ų̀he oyų̀ke ki   ohlate   iyeye.
John letter   that   bed   the   under   found
'John found that letter under the bed.'

This time the literal word-by-word translation of the sentence is hard for an English speaker to understand—an evidence of syntactic difference. It would be very easy, however, for a Japanese speaker to understand. Overall then, we find that most languages have some variation on one of only two basic word orders.

At first glance this is an astonishing result. It would not have been too surprising to discover that all languages had essentially the same word order. This could have happened if human thought had a common underlying logic that determined the form of human languages in some fairly direct way. We would all utter our thoughts in the order that we think them—the order that "makes most sense," as many beginning linguistics students like to think. It would also not have been too surprising to find hundreds or thousands of different word orders realized in the languages of the world, each order uniquely shaped by the culture and traditions of the people who speak the language. Indeed, there is plenty of opportunity for variation. Even if we grant (which we shouldn't) that languages need to have systematic rules governing the orders of their words, there are in principle many systems to choose from. I have described seven generalizations that go a long way toward characterizing word order, and each of these generalizations has two versions, the English version and its reverse. Even if this were all there was to word order, one might expect to find something like $2^7 = 128$ different word order patterns around the world. With only a few more word order principles, there would be enough different possibilities for each of the 6,000 living languages to have its own distinctive order.

But neither of these extremes is what we observe. Rather, we find exactly two common word order types. Putting aside certain minor variations, these two word order patterns account for more than 95 percent of the languages of the world that care about word order at all. Moreover, the two patterns occur in roughly equal numbers. Each type is found on every continent, and each includes more than 40 percent of the languages of the world. (The exact figures depend on controversial details of exactly how one counts languages.) Thus, the word order of human language is not rigidly fixed, but neither does it vary freely. This kind of variation within tightly constrained boundaries is the sign of a parameter at work.

It is useful to think of this situation in terms of the distinction between E-language and I-language. How different are E-English and E-Japanese in terms of word order? Obviously, they are very different: Multiple distinctions in word order magnify and reinforce each other in almost every sentence, posing a formidable challenge to second-language learners. The practical side of this was illustrated vividly for me when I studied Hindi-Urdu with a small group of casual students interested in watching Indian movies and talking to their South Asian spouses. Although Hindi-Urdu is an Indo-European language, distantly related to English, over the centuries it has been influenced by its Dravidian neighbors to the point that it is now an almost pure example of the Japanese-type word order. My class found this extremely confusing. Everyone who did not feel that learning languages was essential to their making a living (everyone but me) dropped out, apparently having decided that they were better off reading subtitles after all. German is a mixed language, with some features of Japanese-style word order. This engenders many jokes among native English speakers about the frustration of waiting until the end of a long German sentence for the main verb to appear.

But we can also evaluate how different English and Japanese are as I-languages. How different are their recipes? Even with no further analysis, the answer would be surprisingly little. The differences boil down to a mere seven discrete statements, all of which can be understood as specifications of how to build the phrases of the language. This is already a tiny number compared to the astronomical number of differences one could enumerate by comparing English and Japanese sentences one by one. Moreover, we can show that the actual I-language difference is even less than this, that only one parameter distinguishes the recipe for English word order from the recipe for Japanese. If all seven differences can be reduced to one fundamental choice, then the fact that there are two predominant word orders in the languages of the world makes perfect sense.

A close look at Table 3.1 provides two hints as to what this basic choice might be. The various elements whose order is fixed are listed in a uniform format in the table. I decided which element to call A and which to call B in such a way that I could always write "A precedes B" in the English column and "A follows B" in the Japanese column. This displays clearly that English and Japanese are opposites in these respects. All the elements in the A column have something in common that distinguishes them from the elements in the B column. The A element is always a simple word: a verb or a noun or a pre/postposition or a "minor-category" word such as complementizer or auxiliary. In contrast, the B element can be more than a single word; it can be a phrase consisting of several words, up to and including an entire clause. This could be a coincidence of these particular examples, but it bears looking into.

The second hint shows up more distinctly in the narrative description of the word order facts, but glimmers of it remain in Table 3.1. The generalization is that the two elements that are ordered with respect to one another also bear a meaning relationship. For example, Table 3.1 states that in English a preposition must come before a *related* noun phrase. This expresses the fact that English has *about the boy,* not **the boy about,* whereas Japanese has the opposite. But it is not true that every preposition in English must come before every noun phrase. On the contrary, the sequence of words *the boy about* is perfectly possible in English, as shown by the following sentence:

I need to speak to *the boy about* his behavior in class.

This is not a counterexample to the statement in Table 3.1 because the noun phrase *the boy* is not directly related in meaning to the preposition *about;* rather, it is related to the preposition *to.* So it must come after *to,* but its order with respect to *about* is undetermined. This is typical. As a second example, consider the statement that auxiliaries come before main verbs in English. This expresses the fact that in English one says *The baby is crying* and not **The baby crying is* (as one would say in Japanese). But again the sequence of words *crying is* is perfectly possible:

The baby's constant *crying is* driving me crazy.

An auxiliary need not come before *all* verbs in English, but it must come before *its* verb, the verb for which it specifies the tense and aspect. The same is true for all the other word order generalizations in Table 3.1: The order of two words is fixed only when they have a certain kind of meaning relationship.

These two hints suggest that the parameter that distinguishes the English-type word order from the Japanese-type has something to do with how words combine with semantically related elements to form phrases. In order to spell out this idea clearly, I need to explain more carefully the notion of a *phrase*.

Phrases are collections of words that appear next to each other in a sentence and are semantically related. They form a single "chunk" of the sentence in terms of both form and meaning. For example, the sequence *a book from her collection* constitutes a phrase in the following sentence:

Kate handed *a book from her collection* to Nicholas.

In terms of word order, these five words are contiguous, one appearing immediately after the other. In terms of meaning, the same five words work together to describe the object that is being transferred in the event described by the sentence. Specifically, the noun *book* says what kind of object it is, the article *a* communicates (among other things) that there is only one of them, and *from her collection* helps to distinguish the particular book from others that may be lying around. Phrases tend to stick together as a unit when the sentence is transformed in some way. Thus, the following two sentences are possible variants of the first; notice that the words *a, book, from, her,* and *collection* are still contiguous:

A book from her collection was handed to Nicholas.
Kate handed to Nicholas *a book from her collection*.

If one tries to split these five words into arbitrary subgroups, the result is gibberish:

*A *from collection* was handed *book her* to Nicholas.
*Kate handed *book her* to Nicholas *a from collection.*

This confirms that phrases are an important aspect of English grammar.

Smaller phrases may be contained inside larger phrases, similar to the way that Russian dolls fit one inside another. This creates a hierarchy of structure between the smallest unit, the word, and the largest unit, the sentence. For example, the noun phrase *a book from her collection* contains as a smaller subpart the prepositional phrase *from her collection.* These three words appear together within the larger noun phrase, and on their own they form a coherent meaning group, specifying an origin. Therefore, they qualify as a phrase in their own right. This prepositional phrase can even be separated from the rest of the noun phrase in some situations, as long as its own integrity is maintained. Thus, we can have:

Kate handed *a book* to Nicholas *from her collection.*

From her collection in turn contains a smaller noun phrase, *her collection,* these two words acting as a unit within the prepositional phrase as a whole. These nested relationships of phrases are found throughout the entire sentence. Linguists often find it helpful to represent these containment relationships in the form of an upside-down tree diagram in which each phrase can be traced up to a single point. Part of the phrase structure of our example, then, could be expressed as in Figure 3.1: Here *S* stands for *sentence,* *NP* for *noun phrase,* and *PP* for *prepositional phrase.* That the words *from her collection* are a linguistically significant unit is expressed in that those words and no others are connected (directly or indirectly) to the node labeled *PP* from the bottom. Similarly, that this prepositional phrase together with the article *a* and the noun *book* form a noun phrase is expressed by the fact that all of these elements are connected to the higher *NP* symbol. The sentence as a whole also counts as a phrase, as shown by the connection of every part of it in one way or another to the top symbol, *S*.

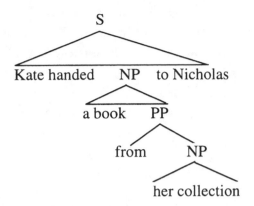

FIGURE 3.1   Nested Phrases in English

The sentences of all languages seem to be built up out of phrases to some degree (although there are significant differences in how this is realized, as we will see in Chapter 4). Phrases exist in Japanese-style languages just as much as they do in English-style languages. The recurrence of expressions like "Xs come before (or after) Ys that are semantically related to them" in my description of word order implies that the key difference between the two language types is found at the level of phrases rather than at the level of whole sentences. After all, phrases are nothing more than groups of semantically related adjacent words. This suggests that the true distinction between Japanese-type and English-type languages involves the details of how words are put together into phrases.

Suppose we think of sentences as being constructed not all at once but in a piecemeal fashion. First one forms a phrase out of one or two simple words taken from the mental dictionary. Then one adds an additional word to that phrase to make a larger phrase. Then another word can be added to form an even larger phrase, and so on. If all goes well, repetitions of this simple procedure eventually result in a complete, well-formed sentence. This view fits well with the highly layered structure that sentences are generally seen to have. But precisely how should a word be added to a phrase to make a new phrase? There are two obvious ways that this can be done, given that speech unfolds in linear time. The new word could be added at the beginning of the already formed phrase, or it could be added at the end. In principle it should make no difference which choice is taken, as long as the speaker and hearer both know what to expect. Japanese-style languages can be seen as those in which new words are systematically added at the end of phrases, whereas English-style languages are those in which words are added at the beginning. Linguists often refer to a word that combines with a phrase to make a larger phrase as the *head* of the new phrase, and they name the whole phrase after it. (Note that this use of the term *head* is similar to, but not exactly the same as, its use in the expression *head-marking language,* introduced in Chapter 2.) For example, *from her collection* is a prepositional phrase, with the preposition *from* as its head, because the phrase was made by combining the single word *from* with the already formed phrase *her collection.* The items that I called "Element A" in Table 3.1 are the heads of their respective phrases within this terminology. The parameter we seek can thus be stated as follows:

THE HEAD DIRECTIONALITY PARAMETER

Heads follow phrases in forming larger phrases (in Japanese,
Lakhota, Basque, Amharic, and certain other languages).
or
Heads precede phrases in forming larger phrases (in English,
Edo, Thai, Zapotec, and certain other languages).

The head directionality parameter subsumes all seven of the explicit generalizations in Table 3.1. The first three rows say that verbs come before direct objects, prepositional phrases, and embedded clauses to which they are semantically related in English, and after them in Japanese. (A direct object is simply a noun phrase that has a direct semantic relationship to the verb.) These are the three kinds of phrases that a verb can combine with in order to form a verb phrase, where a verb phrase is roughly equivalent to the notion of a predicate in traditional grammar. The verb consistently appears on the favored side of this verb phrase—first or last, depending on the language. The next two rows in Table 3.1 mention pre/postpositions combining with noun phrases to make prepositional phrases and nouns combining with prepositional phrases to make noun phrases. We saw examples of both these types of phrases in Figure 3.1. Again, the pre/postposition or the noun head appears consistently on the designated side of its phrase. Other combinations are also possible. For example, a noun can combine with a clause to make a new noun phrase. The noun still comes at the beginning in English:

I heard <u>rumors *that the president will resign.*</u>

(Here the noun phrase is underlined, and the clause it contains is in italics.) Also, a preposition can combine with a prepositional phrase to make a larger prepositional phrase:

A monster emerged <u>from *under the table.*</u>

Phrases like these are also reversed in Japanese.

The last two lines of Table 3.1 are a bit harder to understand because the heads they refer to—complementizers and tense/mood auxiliaries—are less familiar. These are both "minor" grammatical categories. By calling them minor, I don't mean that they are unimportant or too young to vote. A minor category is simply one that contains only a few short and common words. For example, English has approximately four complementizers (*that, if, whether,* and some uses of *for*) and a handful of auxiliaries (*will, would, may, might, can, could, shall, should, must, have,* and *be*). The "major" categories of noun, verb, and adjective, in contrast, contain thousands of members each. These minor-category words are related semantically to the nearby major categories, so we say they *modify* these expressions. For example, the auxiliary *will* adds to *cry* the notion that the shedding of tears takes place in the future. Similarly, complementizers indicate the status of an embedded clause: *That* shows it is finite (it has a tense value), *for* shows it is not finite, *whether* marks it as a question, and *if* marks it as a condition:

> I think that *Chris will win.*
> I would prefer for *Chris to win.*
> I asked whether *Chris would win.*
> I will cry if *Chris wins.*

Since these minor-category words are adjacent to a semantically related verb phrase or clause, they form phrases with them. Since the minor-category words are not phrases themselves, they count as the heads of these new phrases for the head directionality parameter. They thus come before the phrases in English-type languages and after them in Japanese-type languages. Using the terminology defined above, the underlined "embedded clauses" are properly known as *complementizer phrases.* Similarly, *will win* in the first example is an *auxiliary phrase,* made up of an auxiliary head *will* together with the verb phrase *win.*

This last example forces me to clarify what is meant by the term *phrase* in the head directionality parameter. When we merge an auxiliary like *will* with a verb like *win,* we seem to be combining two *words* together, not a word and a phrase. The head directionality parameter as stated does not apply to such a situation. If both expressions are mere words, with the same grammatical status, by what principle does a speaker decide which should come first and which second? Fortunately, the situation is not as problematic as it might seem. Although both partners in this union *may* be single words, the auxiliary can be *at most* a single word. This is not true of the verb. The verb can combine with an object, a clause, or a prepositional phrase to make a verb phrase before combining with an auxiliary, as in *will + [win the 200-meter dash by noon].* The object *the 200-meter dash* is included along with the verb in what *will* says will take place in the future. (Chris might also win the 100-meter dash at the same track meet, but that event could already be in the past at the time we say she will win the 200-meter dash.) It is thus more accurate to state the word order generalization in question as follows:

Modal auxiliaries come before (or after) the related verb *phrase*.

This statement is a special case of the head directionality parameter. We need only add the clarification that a verb, noun, or adjective can constitute a verb phrase, noun phrase, or adjective phrase all by itself. Thus, *win* can count as a verb phrase just as much as *win the 200-meter dash by noon* does, even though it is only one word. In the same way, *pictures* counts as a noun phrase just as much as *beautiful pictures of the Alps*. So when *pictures* is merged with a verb to form a verb phrase, the verb comes first, just as it does when the direct object is more blatantly phrasal:

The magazine will <u>publish *pictures*</u> [*not:* *pictures publish].
The magazine will <u>publish *beautiful pictures of the Alps.*</u>

Not only is the head directionality parameter more elegant than listing the seven specific generalizations in Table 3.1, but it also makes correct predictions about other grammatical constructions. For example, we have not yet considered adjectives. We expect adjectives to combine with prepositional phrases or complementizer phrases to form adjective phrases. If they do, then the head directionality parameter predicts that the adjective should come first in English but last in Japanese. This is correct:

Chris is <u>proud *of the children.*</u>
Julia is <u>happy *that her birthday is coming.*</u>
Hanako-ga   <u>*Taro   yori   kasikoi.*</u>   (Japanese)
Hanako-su   Taro   from   is-smart
'Hanako is smarter than Taro.'

Another class of words in English is the article. This is a minor category, since English has only two true articles (*a* and *the)* plus some related words like *each, no, this,* and *that.* Articles combine semantically with noun phrases to show whether the noun phrase is definite or indefinite. (If I tell you, "I saw the cat," I am assuming that it was a definite cat that you could identify; if I tell you, "I saw a cat," I do not assume this.) The article can be at most a single word, whereas the noun can combine with other elements to form a larger noun phrase. Given this, the head directionality parameter correctly predicts that the article will come before the noun in languages like English and Edo:

> I bought <u>the *picture of Chris.*</u>

It also predicts that articles will come after nouns in Japanese. This is false, but for a rather dull reason: Japanese has no equivalents of the English articles. (Indeed, figuring out exactly when to use *the* and *a* in English is as mysterious to Japanese people as the notorious difference between *l* and *r*.) Other languages with Japanese-style word order, however, do have articles, and in those languages the article follows the noun, as predicted. Here again is an example from Lakhota:

John	wowapi	k'ųhe	oyųke	ki	ohlate	iyeye.
John	letter	that	bed	the	under	found

'John found that letter under the bed.'

The Lakhota equivalent of 'the bed' is *oyųke ki,* literally, 'bed the.' Technically, these expressions, which are normally called noun phrases, should be called article phrases according to our terminology. (Of course, article phrases contain noun phrases, so the distinction between the two terms is not particularly significant.) Thus, the head directionality parameter expresses a pervasive generalization about word order in natural human languages.

Japanese-style word order is not the opposite of English-style word order in all respects, however. Although most expressions are differently positioned in the two languages, there are a few that do not vary.

The most obvious is the subject noun phrase. This comes first in the clause in English, before the auxiliary, the main verb, and whatever else is contained in the verb phrase. The subject also comes first in the normal word order in Japanese, certainly before the verb, and usually before other verb phrase material as well. The normal order of words in English is subject-verb-object, whereas the normal order of words in Japanese is subject-object-verb. The other invariant element is a noun phrase that expresses the possessor of another noun phrase: Such possessors come before the possessed noun in both English and Japanese. Thus, one can create sentences in which the order of words in Japanese matches the English order exactly, as long as the sentence has a verb, a subject, a possessor of the subject, and nothing else:

    John-no   imooto-ga   sinda.
    John-'s   sister-SU   died
    'John's sister died.'

Why are these two configurations immune to the wholesale reversal induced by the head directionality parameter? They are special in that they are both phrases that merge with another phrase, not words that merge with a phrase. As such, they fall outside the domain of the head directionality parameter.

This characterization is clearly true for subjects. The traditional term subject refers to a type of noun *phrase*. Although a subject may be a single noun or pronoun, it can just as well be a complex construction that contains a noun with an article, a prepositional phrase, and so on:

A portrait of Abraham Lincoln should hang on that wall.

The expression that the subject combines with (traditionally called the predicate) is also a complex phrase—typically an auxiliary phrase or a verb phrase. Since the merger of subject and predicate does not involve a head, the head directionality parameter is silent about what its word order should be.

Similarly, the possessor and the expression that describes what is possessed can both be full phrases. For example:

the duke of York's portrait of Abraham Lincoln

In this case the possessor is an article phrase, and the possessed is a noun phrase. Therefore, there is no head, and the order of the combination is also not fixed by the head directionality parameter.

The recipes of both English-style and Japanese-style languages need an additional rule to tell how to build phrases in these special cases. The following rule will do for our purposes:

> When two phrases are combined into a larger phrase and one of them is an article phrase or noun phrase, put the article or noun phrase first.

This holds equally in both kinds of languages.

We might wonder, at this point, whether this phrase-plus-phrase combination rule can also be parametrized. Is this another place where languages have a choice as to how they will proceed? If so, we would expect to find languages that follow the opposite rule:

> When two phrases are combined into a larger phrase and one of them is an article phrase or noun phrase, put the article or noun phrase last.

If this rule were used in a head-first language, the result would be a language in which word order is overall similar to English but subjects come at the end of sentences and possessors come at the end of noun phrases. If on the contrary this rule were used in a head-final language, then the result would be true Mirror English, with words coming in the exact reverse of the order found in English.

It is not entirely clear whether such languages exist or not. If they do, they are rare. There are a few languages in which subjects come at the end of sentences, but they are a small minority, somewhere between 1 percent and 3 percent of languages of the world. Tzotzil, a Mayan language spoken in southern Mexico, is an example.

```
7i-yal-la        ta       te7     ti    vinik-e.
PAST-descend     from     tree    the   man
'The man came down from the tree.'
```

(The numeral 7, used twice in this example, stands for a glottal stop, a consonant not found in English.) Apart from the placement of the subject, this word order is very much like English, with a tense particle *(7i)* before the verb, verb before prepositional phrase, preposition before noun phrase, and article before noun. Possessors also come after the possessed noun in Tzotzil, as expected:

```
stzek      li       7antze
skirt      the      woman
'the woman's skirt'
```

Some Austronesian languages, including Malagasy, the language of Madagascar, also have this word order. The Mirror English word order is even rarer, with only a handful of attestations, all from little-known Carib languages of South America, such as Hixkaryana:

```
Kanawa  yano    toto.
canoe   took    person
'The man took the canoe.'
```

It is somewhat peculiar that there are so few subject-last languages, whereas head-first and head-last languages exist in roughly equal numbers. There are also some signs that these word orders are not completely consistent, particularly in Hixkaryana. In Chapter 6 I return to some of these details, together with a high-tech alternative analysis that has been put forward recently. For the time being, however, I tentatively assume that this is a real parameter, particularly relevant to Tzotzil and Malagasy. I call it the *subject side parameter.*

Returning to our primary comparison between Japanese and English, we have learned that in terms of I-language they are very similar. They enter the process of sentence construction with similar repertoires: Both have nouns, verbs, adjectives, pre/postpositions, complementizers, and auxiliaries. Both types of languages build up sentences in piecemeal fashion, by joining words into phrases that are understood as units of meaning. Both languages follow the same rule for combining two phrases into one larger phrase. Both follow a consistent rule for adding a word to a phrase. The one difference has to do with which side the extra word is added on. Thus, the difference in the recipe for word order in Japanese as opposed to English is scant. Nevertheless, this very small difference in the recipe has a huge impact on the E-language—on the kinds of sentences that are actually formed—simply because each language uses its version of the rule many times in the formation of a typical sentence. This illustrates my point that languages are basically recipes for making sentences, and parameters are atomic instructions in those recipes where languages have a choice.

The similarities and differences between English and Japanese can be expressed visually in a way that some readers may find helpful. Earlier I described a kind of inverted tree diagram, in which all the words of a phrase are connected by lines (directly or indirectly) to a single node that stands for that phrase. Figure 3.2 shows such a diagram for an ordinary English sentence, alongside one for a Japanese sentence of similar structure and complexity. (S stands for a clause, the result of merging a subject with an auxiliary phrase.) These diagrams again show us the vast differences in word order between English and Japanese. But they also show a deep similarity. Imagine that these tree diagrams are really Alexander Calder mobiles, with the lines made of strong wire and the words made out of metal sheets. Imagine further that they are attached to the ceiling by the S and allowed to twist and swivel freely in a light breeze. Then you will realize that the Japanese mobile and the English mobile have exactly the same design. The only difference between the two is that every

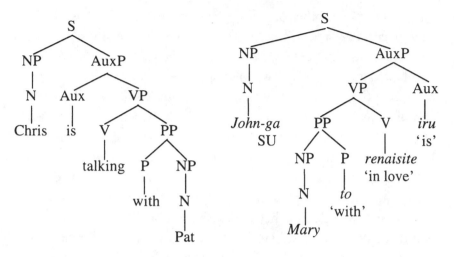

FIGURE 3.2   English and Japanese Phrase Structure Compared

node whose label ends in *P*—every phrase consisting of a word joined with another phrase—has swiveled around in the English version relative to its position in the Japanese sentence. Only the top *S* node, which consists of two full phrases, has not swiveled. (If it should swivel, too, the result would be Tzotzil, which is Mirror Japanese.) The word order in English and Japanese is very different, but the overall geometry of the sentence is exactly the same.

———————

One can agree with everything I have said so far and still disagree strongly about *exactly* what a language is, and therefore have different views about *exactly* what a parameter is. Linguists argue about these basic notions just as biologists who agree on a general theory of evolution nevertheless have different ideas about exactly what a gene is or how natural selection operates. Some linguists avoid using the term *parameter* at all, preferring other terms with somewhat different connotations. Still, I believe that there is a basic notion in common underneath the disagreements about details.

The original way of looking at parameters is to say that a (syntactic) parameter is a rule of sentence formation that holds in one language but not in another or a rule that exists in two versions, of which a language must adopt one. An alternative view, proposed by Alan Prince and Paul Smolensky, is that exactly the same rules are operative in all languages. On this view, languages differ not in which rules they have but in the priorities they assign to the universal stock of rules. For example, this kind of theorist would say that the rule that heads come before phrases (as in English) *and* the rule that heads come after phrases (as in Japanese) both apply to all languages. The two rules contradict each other, so it is usually impossible to obey both perfectly when constructing a phrase. The difference between English and Japanese is in which rule they consider to be the more important, the head-first rule or the head-last rule. On this view, a grammatical sentence in (say) Edo is not one that obeys the rules of Edo perfectly (as I have assumed so far) but one that best respects the priorities of Edo in the midst of the various conflicting requirements of grammar. This approach is therefore known as optimality theory.

In ordinary situations of word order, optimality theory does not make different predictions from the more standard theory, but interesting differences can arise in special cases. For example, suppose one had a language in which heads typically came at the beginning of phrases (as in English), but some special circumstance intervened to make this impossible in a particular situation. Optimality theory might predict that since the priority of having the head at the beginning cannot be fulfilled, the lower priority of having a head at the end would rise to the top of the priority list. Head finality is never expected in the classical approach: If a head cannot come at the beginning for some reason, one would expect it to get as close to the beginning as it could. A possible case in support of the optimality theory view comes from the African language Nupe. In Nupe complementizers usually come before the clauses they introduce, as in English:

Mi	kpaye	gànán	*Musa*	*lá*	*èbi.*
I	think	COMP	Musa	took	knife

'I think that Musa took the knife.'

Focus constructions, however, work differently. In these constructions a noun phrase expressing important new information is placed in the most salient position, at the beginning of the sentence. When this happens, the complementizer loses its chance to come first in the clause. Rather than settle for second place, it comes at the very end of the clause:

Èbi   Musa   lá   o.
Knife Musa   take COMP
'[It's] a knife that Musa took.'

As an instance of head finality in an otherwise head-initial language, this seems to support the view that both rules are active in the language, even though the one with higher priority normally conceals the effects of the one with lower priority. Optimality theory has become the dominant view about how languages differ in the study of phonology, and some people are using it to study sentence structure as well.

Another variant view is that parameters are all keyed to the basic properties of certain words. This lexical parameter approach was originally proposed by Hagit Borer and has been developed by people like Naoki Fukui and Chomsky. Part of its motivation comes from language acquisition. As a bare minimum, learning a language must involve learning the language's distinctive words. There is no escape from this: It is impossible to deduce from general principles of cognition that a domestic canine is called *dog* in English, *perro* in Spanish, *erhar* in Mohawk, and *èkítà* in Edo. Since this kind of learning must happen in any case, it would be nice if learning the grammatical properties of a language somehow piggybacked on the task of learning the words. And it is true that in many cases the special syntactic properties of a language seem to be determined by particular words or groups of words. For example, theorists of this persuasion would look at the Nupe sentences above and focus on the fact that the head-initial complementizer in the first sentence is not the same as the head-final complementizer in the second sentence: One is *gànán* and the other is *o*. A child learning Nupe must learn that the complementizer in an ordinary embedded sentence is pronounced one way and the one in a focus sentence is pronounced another way. Then the lexical theorist can say that the child simply learns which complementizer comes before the clause and which comes after it at the same time as learning the pronunciations. This theory needs to be elaborated in some way to explain why virtually every head in English comes before the associated phrase on its right, whereas virtually every head in Japanese comes after its phrase. We do not find languages in which the verb meaning 'hit' comes before the object, English-style ("The child might hit his parent"), and the verb meaning 'kiss' comes after it, Japanese-style ("The child might his parent kiss"). The word order of the object and the verb is thus not learned purely by learning 'hit' and 'kiss' as individual verbs but must be somehow keyed into the process of learning the verbs as a class. If this concern can be worked out, however, the lexical parameter approach has some attractive properties.

Perhaps the deepest divide in how linguists think about linguistic diversity is the one between the formalists and the functionalists. These two approaches do not disagree so much about the facts of linguistic diversity—both, for example, believe in Greenberg's word order universals; where they differ most is in how they explain these universals.

Formalists typically seek what are sometimes called *internal* explanations, where one feature of a language is explained in terms of another feature of the same language. Often this works itself out as a claim that the same general rule or principle is determining both features. The account I sketched of the word order differences between English and Japanese is an internal, formalist explanation in this sense.

Functionalists, in contrast, prefer *external* explanations, in which features of languages are explained in terms of other aspects of human cognition that do not necessarily have anything to do with language. Functionalists are not as likely to appeal to abstract general rules. They do not feel compelled, for example, to say that there is a single rule that applies to both prepositional phrases and verb phrases. Instead, they are inclined to say that languages with preposition-object order are also likely to have verb-object order because that makes the language somehow more efficient for people to parse or produce, given general properties of human cognition. Because languages that have object-postposition order and object-verb order are also efficient to parse and produce, there is no functional pressure for a language to prefer one system over another, so long as it is consistent. Yet languages that mix, say, object-postposition order with verb-object order cannot be computed so easily, they argue. Such languages could exist, but they would be rare and prone to being "simplified" over time into one of the more stable and efficient systems.

Functional linguists also seek to understand parameters in terms of history. For example, in many West African languages and in East Asian languages like Chinese and Vietnamese, prepositions have evolved historically from verbs. For example, the Nupe word *ya* 'for' comes from the verb meaning 'give,' and the Nupe word *be* 'with' comes from a verb meaning 'come.' Since verbs come before objects in Nupe, and since prepositions have evolved from verbs, it is not surprising that prepositions also come before their objects. Functionalists would say that this gross word order is simply carried over from the earlier stage of evolution.

Fortunately, we do not have to decide right now which of these approaches is correct. We have plenty to do in understanding the patterns of linguistic variation without facing all of the ultimate why questions directly. Moreover, there is no logical conflict between the two modes of explanation. Both may be essentially correct at different levels of description. Still, formalist-style explanations have one major advantage for a book like this: They allow me to stay focused on the facts of human language itself without having to introduce a great deal of background information about other aspects of human cognition. In fact, most linguists have a limited background in general cognition, which makes us vulnerable to saying things that are silly or full of wishful thinking when in search of an external explanation. For this reason, I present parameters in a broadly formalist idiom, though not necessarily denying the possibility of other levels of description and explanation.

———————

Functionalists also differ from formalists in another way: They tend to see continuous variation across languages, whereas formalists tend to see languages as falling into discrete types. My portrayal of word order was formalist in spirit in this respect, too. I presented two distinct types of word order, with no intermediate cases or transitional forms. Functionalists would rightly point out that this is a bit of a simplification. I already mentioned that there are a few head-final word orders in Nupe and German, languages that are otherwise head-initial. Another example is Amharic, which has verbs that come after their objects but has prepositions rather than postpositions. It thus offers a kind of blend between Japanese-style word order and English-style word order:

Aster   wənbər-u-n   lə-setiyyə-wa   sət't'əčiw.
Aster   chair-the-OB to-woman-the   gave
'Aster gave the chair to the woman.'

Such examples notwithstanding, on this point I have to disagree with the functionalists more strenuously. It is true that the formalist's discrete types are to some extent idealizations away from the noise in actual human languages. But they are no more extreme idealizations than those of classical physicists, who said that objects fall to earth at a rate of 9.8 meters per second squared. Very few objects really fall at exactly this rate; certainly the leaves shed from the trees in my backyard do not. Nevertheless, many objects fall at close to this rate, and the more we control for other factors (such as air resistance), the closer they approach the theoretical value. The same seems to be true for languages: Languages that are close to the ideal types are much more common than languages that are far from them. According to the statistics of Matthew Dryer, only 6 percent of languages that are generally verb final are like Amharic in having prepositions rather than postpositions. Moreover, although practical and ethical considerations prevent linguists from building the equivalent of vacuum tubes to see what language people would learn under carefully controlled conditions, it does seem that the noise we see can often be attributed to other factors. Amharic, for example, is a Semitic language, related to pure head-initial languages like Arabic and Hebrew but spoken in Africa, in the environment of head-final Cushitic and Nilo-Saharan languages. The conflict of historical and geographical influences could partially explain why Amharic is a mixed case.

Meanwhile, the real weakness of the "continuous variation" view is that it has a hard time explaining why many logically possible forms of language are extremely rare or nonexistent. However you look at it, the range of possible human languages is much smaller than the range of conceivable ones. There may be more than two word order types, but there are still many fewer than the 512 (or more) possibilities predicted by simple combinatorics. The same is true for the null subject parameter discussed in the previous chapter. Strictly speaking, this parameter predicts that there should be only two kinds of languages: perfect French/English and perfect Italian/Spanish. In fact, when one looks carefully at southern French and northern Italian dialects, one finds a few transitional forms and mixed cases that cloud the picture. But these forms are rare, and they still add up to only a fraction of the logical possibilities. When Gary Gilligan looked at the four distinctive features related to the null subject parameter as outlined in Chapter 2, he found that his sample of languages contained just seven combinations that occurred more than once and two more that were attested only once. This is a bit more than the three to five language types that the parametric theories of his day predicted would exist. But it is also a long way from the continuous variation that an "anything goes" theory would predict. The four distinctive features could logically be combined into sixteen (2^4) different language types, but roughly half of those types were never observed, even in a relatively superficial study that did not attempt to filter out any noise. It is certainly not the case that anything goes. Although we may very well need to include additional, finer-grained parameters to account for the "extra" languages, we would not be well advised to abandon discrete parameters to embrace continuous variation.

In short, those who have been skeptical of the notion of a parameter because it does not seem to do justice to the continuities that can sometimes be observed in human language fall into the same class as the chemists of Dalton's day who could not accept the idea that matter is made up of discrete particles. Our normal observations of continuous variation are undeniably persuasive. But for chemistry, the solution was not to abandon the idea of discrete atoms but to find more of them. Similarly, the way forward for linguistics is not to abandon parameters but to find more of them, enriching and developing the theory until it really does account for the full range of variation.

4

Baking a
Polysynthetic Language

SUPPOSE THERE WAS A CONTEST for the language whose grammar was the least like English. Probably no one language would emerge as the clear winner. Languages differ in various ways, and (as with other beauty contests) the selection is partly a matter of the judges' personal tastes. Nevertheless, the Mohawk language would be a serious contender.

Mohawk is spoken by the Native Americans who were the guardians of the eastern door in the once powerful Iroquoian Confederacy in upstate New York. It is still used by some 2,000–4,000 people in Quebec, Ontario, and New York. Mohawk is what linguists call a *polysynthetic* language. Essentially, this means that words in Mohawk are extremely long and complex, typically made up of many distinct parts. Sentences, in contrast, are relatively short and fluid compared to English; they can consist of just a single verb. When additional noun phrases appear, they can be arranged around the verb in practically any order one likes. Thus, Mohawk fits neither of the two common word order types described in the previous chapter. The parts of a noun phrase do not even always appear next to each other in Mohawk, calling into question whether sentences are built up out of phrases at all. The cumulative effect of these features is to make

Mohawk look extremely different from English. The army, in fact, experimented with using Oneidas (cousins of the Mohawks, whose language is almost the same) as Code Talkers in World War II. If they had got around to implementing this plan more vigorously, Mohawks might have won the same fame that the Navajos did.

Winning a conventional beauty contest can lead to new opportunities: appearing at charity functions, endorsing household products, perhaps even an acting audition. By winning the "different from English" crown, Mohawk also gains a new opportunity: the privilege of being a testing ground for the ideas we have been exploring about language and parameters. So far, in introducing parameters and explaining why they were proposed, I have focused on the easy cases. Null subject phenomena and word order patterns were the instant successes that got parametric reasoning planted in linguists' minds in the first place. But a scientific theory cannot thrive on easy cases alone. The ultimate test of a theory is not whether it accounts for the facts that initially motivated it—any hypothesis should be able to do that—but whether it can be applied to the hardest cases, those that played no role in inspiring the theory and that looked at first sight like they would be recalcitrant. If a theory also works in these situations, then it is time to take it seriously as an expression of something real and general. Mohawk offers just this kind of proving ground for the parametric theory of linguistic diversity.

It was with this in mind that I began a five-year study of the Mohawk language, seeking to put ideas about universal grammar and parameters to the test. (Mohawk also happened to be the most interesting language that I could study in its native environment and still get home in time to put the children to bed.) As we saw with word order in Chapter 3, the challenge is to tease apart whether or not the extensive differences between E-Mohawk and E-English show that there are extensive differences between the I-languages. Are Mohawk and English incommensurable grammatical systems, with so many basic differences that they cannot be meaningfully compared with each other? In this chapter I try to convince you that the answer to this question is no.

In fact, it turns out that I-Mohawk differs from I-English in one relatively small way, but the difference is strategically placed to have a huge effect. In other words, the English system and the Mohawk system are actually distinguished by a single parameter. In terms of the cooking analogy I used earlier, the recipes for the two languages are essentially the same except for one ingredient: Mohawk is English with a tablespoon of yeast added. It is in that sense that I show how to "bake" a polysynthetic language in this chapter. This is perforce my most extended single example of parametric reasoning in the book. But the reader who is willing to follow the details will be rewarded with a striking illustration of the power and utility of this kind of theory for understanding even the most extreme differences among languages.

As a first step, we should appreciate Mohawk on its own terms.

The term *polysynthetic* was coined by the early typologists of the nineteenth century. Languages were called *synthetic* if they tended to express grammatical relationships by changing the forms of words. Classical Latin and Greek were the classic examples of synthetic languages: Learning Latin and Greek involves mastering a seemingly endless set of case declensions (for nouns) and tense conjugations (for verbs). A polysynthetic language is one that is synthetic to an extreme degree, a language that expresses nearly all grammatical relationships by elaborations on the verb, beyond what one finds in any Indo-European languages. In practice, this means that words in Mohawk can be extremely long and complex, expressing the equivalent of whole sentences in English. One of my favorite (though slightly artificial) examples is:

Washakotya'tawitsherahetkvhta'se'.
'He made the thing that one puts on one's body [i.e., the dress] ugly
    for her.'

This is a thirty-three-letter word, one shorter than *supercalifragilisticexpialidocious,* Mary Poppins's famous sign of English verbal skill.

(Note that the apostrophes in this example are genuine letters in Mohawk; they represent a so-called glottal stop, a sound not found in English but common in other languages.) Unlike *supercalifragilistic-expialidocious,* however, the Mohawk word actually means something, the most literal English translation of which needs a thirteen-word sentence.

A word like this might give you nightmares about what vocabulary tests must be like in a Mohawk 101 class. But even the native Mohawks do not learn such words. Rather, they make them up and understand them on the fly. This feat is somewhat comparable to the ability of English speakers to form new compounds and understand them immediately, which is something that we do all the time. For example, you will not find *turkey-stranglers* listed in a dictionary, but if I used this word you would immediately recognize it as English and know what it meant. A Mohawk speaker's use of *washakotya'tawit-sherahetkvhta'se'* would be similar. The example is still intimidating, though, since the word consists of at least eleven meaningful parts, and they must be arranged in exactly the right order. Any juggling of these parts would result in meaningless gibberish. This aspect of Mohawk is very complicated from the English perspective. And my illustrative example is only a slight exaggeration from the Mohawk norm. One of the first verbs in the Mohawk teaching grammar is *katerihwaiénstha',* which literally means something like 'I habitually cause myself to have ideas,' or more idiomatically, 'I am a student.' There are no short, simple words one can use to ease into the task of learning Mohawk.

Polysynthetic languages are also *nonconfigurational,* meaning that sentence structure seems very loose and fluid in comparison to English. In English a transitive verb such as *like* needs to appear with two noun phrases, a subject and an object, in order to make a complete, well-formed sentence. Furthermore, the arrangement of these noun phrases is more or less fixed: The subject (or liker) must come before the verb, and the object (or liked thing) must come immediately after it. Mohawk contains neither these syntactic rules nor their Japanese-like mirror images. The word-for-word translation of a simple English sentence is grammatical in Mohawk, too:

Sak	ranuhwe's	ne	atya'tawi.
Sak	likes	the	dress.

But there is nothing special about this arrangement; any other ordering of the same three content words is just as good, depending on which part of the sentence the speaker wants to emphasize. Thus, in Mohawk one can also say any of the following:

Ranuhwe's	ne	atya'tawi	ne	Sak.
Likes	the	dress	(the)	Sak

Ranuhwe's	ne	Sak	ne	atya'tawi.
Likes	(the)	Sak	the	dress

Sak	atya'tawi	ranuhwe's.
Sak	dress	likes

Atya'tawi	Sak	ranuhwe's.
Dress	Sak	likes

Atya'tawi	ranuhwe's	ne	Sak.
Dress	likes	the	Sak

One can also leave out either the subject noun phrase or the object noun phrase or both, while putting the remaining words in any order. The following thus also count as complete and natural Mohawk sentences.

Ranuhwe's	ne	atya'tawi	[or *Atya'tawi ranuhwe's*].
Likes	the	dress	[i.e., 'He likes the dress.']

Sak	ranuhwe's	[or *Ranuhwe's ne Sak*].
Sak	likes	[i.e., 'Sak likes it.']

Ranuhwe's.	
Likes	[i.e., 'He likes it.']

English, of course, does not tolerate such freedom in sentence formation.

Overall, Mohawk seems to be a very different linguistic system from English, complex and rigid where English is simple but simple and free where English is complex. Languages like Mohawk thus pose a serious problem for the idea of an innate, universal grammar—what Pinker calls "the language instinct." On the one hand, if children are born with an instinctive knowledge of the English patterns, it seems that this will not help them learn Mohawk. On the other hand, if children are born with an instinctive knowledge of the Mohawk patterns, this will not help them learn English. Yet children who grew up in Kahnawake, Quebec, where I studied Mohawk while teaching at McGill University in Montreal, successfully learned both.

———————

As the first step toward understanding this particular version of the Code Talker paradox, let us look more closely at exactly what is involved in making Mohawk polysynthetic. What is going on inside those long and complex words?

One of Mohawk's most important polysynthetic features is called *noun incorporation*. Noun incorporation involves adding together a noun and a verb to create a single new word, the meaning of which includes the meanings of the two parts. The thirty-three-letter Mohawk word that I offered earlier contained an instance. That word combines the noun *atya'tawi,* literally, 'thing one puts on one's body' (the normal Mohawk expression for a dress or a shirt—any garment that covers the torso), with the verb *hetkvht,* meaning 'to make something ugly.' A less fanciful example of noun incorporation is:

Owira'a   wahrake'   ne   o'wahru.   (plain version)
Baby   ate   the   meat

Owira'a   waha'wahrake'.   (noun incorporation version)
Baby   meat-ate

The first of these sentences is comparable to English; it has a subject, a verb, and an object, each expressed by a separate word. In the second sentence, the core of the word 'meat' *('wahr)* combines with the core of the word 'eat' *(k)* to give a larger word *'wahrak,* 'to eat meat.' (All of these words then take additional prefixes and suffixes to express tense and agreement, in ways that I come back to.) Such noun incorporation is common in Mohawk. Some other typical examples are:

Wa'eksohare'.	'She dish-washed.'	*(ks* 'dish' + *ohare* 'wash')
Wa'kenaktahninu'.	'I bed-bought.'	*(nakt* 'bed' + *a* + *hninu* 'buy')
Wahana'tarakwetare'.	'He bread-cut.'	*(na'tar* 'bread' + *a* + *kwetar* 'cut')

(Notice that if the noun stem ends in a consonant and the verb stem begins in one, the vowel *a* is inserted between the two parts. This helps to prevent pileups of consonants, making the words easier to articulate.)

At first this looks like an excellent illustration of how different languages can be. After all, English speakers cannot say **The baby meat-ate* or **He finally bread-cut*. These Mohawk-imitating examples are reasonably understandable; if someone were to utter one, you would know what the speaker meant. But people ordinarily do not utter them, and if they did, the spell-checkers and copyeditors of the world would hurry to correct them. Noun incorporation simply is not part of English the way it is of Mohawk.

Nevertheless, English does have something rather like noun incorporation in a slightly different domain. Whereas one cannot generally combine a noun and a tensed verb in English, one can combine a noun with another noun to form compounds like *doghouse* or *earwax*. Moreover, the second noun in such a compound can be derived from a verb by adding a noun-forming suffix such as *–er* or *–ing*. Thus, even though it is bad to say **The husband dishwashed*, it is fine to use *dishwashing* and *dishwashers*. Also, like the Mohawks, English speakers can easily make up new words built on this pattern: A particular kind of furniture store employee, for example, could be called a *bed-buyer*. So Mohawk and English are actually similar in that both have productive compounding. They differ only in that Mohawk can create noun-plus-verb compounds whereas English is limited to noun-plus-noun compounds.

Noun incorporation in Mohawk and compounding in English are also similar in a more subtle way. Think for a moment about an act of washing. Two characters are necessarily involved in such an act: an agent, which is expressed in English as the subject of the sentence, and an undergoer, which is expressed as the object. These two participants show up as separate phrases in an ordinary sentence:

The husband washed the dishes.

Now, it is easy for the object *dish* to appear in a compound with the verb, once the verb has been made into a noun. One can construct natural-sounding examples such as:

The husband enjoys dishwashing.
The husband is a good dishwasher.

Suppose, however, that one tries to make similar compounds out of a nominalized verb and a noun that is supposed to be understood as the subject of the verb, the doer of the action. Basic fairness suggests that this should also be possible. But such combinations are systematically impossible in English:

*She appreciates husband-washing [of dishes].
*He is a good husband-washer [of dishes].

This pattern is very consistent in English: One can say *bread-cutting* (where bread is the object of cutting) but not **knife-cutting (of bread)* (where knife is the agent of cutting). One can say *meat-eater* but not **dinosaur-eater (of meat)*, and so on. This can even be seen in the novel compound *turkey-strangler*, which I coined above and predicted you would understand. Doubtless you saw that the turkey would be the victim of the strangling and not the perpetrator, despite your having no prior experience with the word and no context from which to infer this.

Exactly the same "unfairness" holds of noun incorporation in Mohawk. A noun that expresses the object of the verb can be combined with the verb root to make a larger word, as in all the examples that we have seen so far. But similar-looking combinations of a verb with a noun that expresses the subject of the verb (the agent of the action) are impossible. The ordinary Mohawk sentence *Owira'a wahrake' ne o'wahru* (Baby ate the meat) can therefore very well correspond to a sentence with the object incorporated, as we saw earlier. But it cannot correspond to a sentence with the subject noun incorporated:

    *Wahawirake' ne   o'wahru.
    Baby-ate   the   meat

Nor does changing the word order in this sentence improve it. Similarly, one cannot say *Wawahsarakwetare' ne kana'taru* (knife-cut the bread) to mean 'A knife cut the bread.' Mohawks who hear such an utterance generally laugh and tell you that bread cannot cut a knife. This reaction shows that they *always* understand the incorporated noun as the object of the verb—even when that flies in the face of our shared knowledge about what kind of events usually involve knives, bread, and cutting.

This notable parallel in the relationships a compound can and cannot express shows even more clearly that Mohawk and English are not as different as we might have thought. The similarity can be expressed in the following principle of universal grammar, which I'll nickname the "verb-object constraint":

THE VERB-OBJECT CONSTRAINT

The object of a verb must be the first noun (phrase) to combine with the verb; the subject cannot combine with the verb until after the object does.

The verb-object constraint was first proposed exclusively for English and related languages. It explains not only the facts about compounds that we just reviewed but also the placement of subjects and objects in simple English sentences. We saw in Chapter 3 that objects come immediately after the verb in English, whereas subjects come before it. Moreover, subjects do not necessarily come *immediately* before the verb: They can be separated from it by certain other elements, such as adverbs and auxiliaries (e.g., <u>Chris</u> *will quickly* <u>*wash*</u> *the dishes*). For this and other reasons, we said that the object noun phrase is joined together with the verb to form a verb phrase, whereas the verb and the subject do not form any unit other than the sentence as a whole (see Figure 3.2). In short, the object must be inside the verb phrase whereas the subject cannot be, just as the object may be inside a compound but the subject cannot be. These two observations are expressed in a consistent way in the verb-object constraint. In Chapter 3 I took it for granted that this same constraint also holds in Japanese-style languages. Japanese differs from English in that the verb is added after rather than before a noun phrase to form a verb phrase. The noun phrase that the verb joins with, however, must be the object, not the subject in both languages. In Japanese this results in the object's appearing immediately before the verb, between it and the subject:

> John-ga   *tegami-o*   *yonda.*
> John-SU   letter-OB   read
> 'John read the letter.'

The behavior of noun incorporation shows us that Mohawk words obey this same verb-object constraint—as do complex words in all the other polysynthetic languages we know of. The verb-object constraint thus qualifies as part of universal grammar, its effects being seen in all languages in one way or another.

These facts also show something else even more basic: Mohawk must be like English and Japanese in distinguishing subjects from objects in the first place. Furthermore, the agent of the action counts as the subject and the undergoer of the action counts as the object in all three types of languages. (This is not to be taken for granted, as we will see in Chapter 6.) If these things were not so, then the verb-object constraint either would not apply to Mohawk at all or it would give a very different pattern of facts. Thus, this whole cluster of concepts transcends the distinction between polysynthetic and non-polysynthetic languages.

So far, then, the only difference between Mohawk and English we have found amid all these similarities is that Mohawk allows nouns to compound with verbs and English does not. Exactly what this special opportunity in Mohawk can and cannot be used to express is determined by known principles of universal grammar.

There is much more to polysynthesis than just noun incorporation. But many other aspects of polysynthesis can be understood in more or less the same way. For example, another ingredient in my original example, 'he made the thing that one puts on one's body ugly for her,' is the so-called causative suffix. Normally *hetkv* 'to be ugly' is an intransitive verb in Mohawk—meaning that it appears with a subject but no object. This reflects that being ugly is a one-person job; it is something you can do alone by yourself in your room. (Perhaps that is the best way to do it.) The suffix –*ht*, however, can attach to this verb root to make a transitive verb, *hetkvht* 'to make something ugly.' This new action involves two participants: the thing that becomes ugly and the thing that makes it ugly. The same causative suffix can also be seen in this simpler example:

Ashare'   tu*h*sv'ne'.
Knife   fell-down
'The knife fell.'

Uwari tayú*h*s*v*hte'   ne   áshare'.
Mary made-to-fall   the   knife
'Mary made the knife fall.'

The first sentence here shows the ordinary intransitive verb *hsv* 'fall'; the second one shows this verb combined with -*ht* to give a transitive verb *hsvht* 'make fall.' In English these causative events are typically expressed by combining two distinct verbs like *make* and *fall* in a special syntactic configuration, but Mohawk uses a single complex word to express them. This, then, is another facet of Mohawk's polysynthesis.

Nevertheless, here, too, we find not only differences but also similarities to more familiar phenomena. The Mohawk causative verbs are similar to noun incorporation in that they combine into a single complex word two elements that would normally constitute separate words in English. The only difference is that this time both of the elements (*hsv* 'fall' and *ht* 'make') have the meaning of verbs, not nouns. So we could call this verb incorporation. Moreover, something like the verb-object constraint applies in these examples as well.

To see this, let's look more closely at the English word *make*. It can be an ordinary transitive verb, which takes an object noun phrase, as in *I made <u>a necklace.</u>* This sentence means something like 'a necklace exists as a result of something I did.' Instead of taking a noun phrase object, however, *make* can merge with a special kind of verb phrase to create a sentence like *I made <u>the knife fall.</u>* The meaning of this sentence is somewhat similar to the meaning of the first one: It can be paraphrased as 'an event of the knife's falling exists as a result of something I did.' We can extend traditional usage slightly to say that the direct object of *make* is the verb phrase containing *fall* in this second sentence. The only real difference is that noun phrase objects typically refer to things, whereas verb phrase objects refer to events.

Just as *I made a necklace* and *I made the knife fall* are parallel syntactic formations in English, so noun incorporation and causative verbs are parallel word formations in Mohawk. In all four cases, the verb-object constraint is respected: The object of the verb (whether noun or verb) combines with the main verb (making either a phrase or a complex word), but the subject does not. Similar formations are sometimes found in English. There is no common one-word equivalent to 'make something ugly' or 'make something fall' in normal English, but we do have words like *beautify,* which means 'to make something beautiful,' and *modernize,* which means 'to make something modern.'

(The root words here are adjectives in English, whereas they are verbs in Mohawk, but the basic pattern is the same.) Furthermore, in C. S. Lewis's children's story *The Voyage of the Dawn Treader*, the Dufflepods speak of having been "uglified." This expression is intended as a kind of joke, but it is an easily understandable one.

Using these examples, we can conclude that polysynthesis in Mohawk is not completely wild and woolly. It obeys the same general rules of linguistic construction that words and phrases do in other languages. Also, partially similar formations are found in English. Polysynthesis is the result of using the same kinds of word formation procedures found in English, but more frequently and in more varied situations. It is a difference in degree rather than a difference in kind—a variation in the recipe for a natural language rather than a whole new recipe. Thus, there does seem to be a parameter at work here. But before we can discern what kind of parameter it is, we must consider Mohawk's nonconfigurational properties more carefully.

Recall that the second class of differences between Mohawk and more familiar languages is that Mohawk sentence structure is very fluid, with subject and object appearing anywhere in the sentence or even omitted entirely. All of the following sentences, then, are possible:

| Atya'tawi | Sak | ranuhwe's. | | |
| Dress | Sak | likes | | 'Sak likes the dress.' |

| Ranuhwe's | Sak | ne | atya'tawi. | |
| Likes | Sak | the | dress. | 'Sak likes the dress.' |

| Sak | ranuhwe's. | | |
| Sak | likes | | 'Sak likes it.' |

Notice that in the first two examples the subject is closer to the verb than the object is. It seems that there could be no phrase that contains the verb and the object but not the subject in these examples, in violation of the verb-object constraint. Thus, although this constraint restricts the forming of words in Mohawk, it does not seem to restrict the forming of sentences the way it does in English-type and Japanese-type languages. The third sentence, in contrast, seems to violate a very general completeness condition, which says that a transitive verb that expresses an event involving two participants must appear with two noun phrases, one to denote each participant. This completeness condition implies that in English we cannot say *Chris likes* or *Pat devoured*. Mohawk is apparently different in this regard, too.

Yet we have seen patterns like this before, when discussing the null subject parameter in Italian and Spanish back in Chapter 2. Recall that Italian and Spanish (unlike French or English) allow the subject of a tensed sentence to be "inverted" to the end of the sentence or to be left out entirely:

> Gianni   verrà.
> Gianni   will-come
> *or*   Verrà   Gianni.
> Will-come Gianni
> *or*   Verrà.
> [He] will-come.

Perhaps Mohawk is similar, except that the freedom that the subject enjoys in Italian and Spanish is extended to the object noun phrase as well. If so, we could say that Mohawk is a null subject and object language. Once again, Mohawk would be different from less exotic languages in degree rather than kind.

There is a rather obvious functional reason why subjects can be left out in Spanish and Italian but not in French and English, which I did not mention in Chapter 2 but will prove helpful here. In Italian and Spanish, verbs take different endings depending on what their subject is. A typical set of examples from Spanish is:

> yo com*o*   'I eat'
> tu com*es*   'you eat'

el com*e*	'he eats'
nosotros com*emos*	'we eat'
vosotros com*eís*	'you all eat'
ellos com*en*	'they eat'

The basic verb root *com* 'eat' thus takes different suffixes for first person, second person, and so on. As a result, subject pronouns in Spanish are redundant because the intended pronoun is conveyed by the verb ending. It is not surprising, therefore, that the subject can be left out in circumstances where a simple pronoun would be used as a subject in English. French and English verbs used to be like Spanish and Italian in this respect, but most of the verb endings were lost as the sound patterns of the languages were simplified. (The historical endings are still used in standard written French, but many of them are no longer pronounced.) Thus, subject pronouns in French and English are not redundant and must be retained.

In Mohawk, objects as well as subjects can be omitted. Moreover, omitting either a subject or an object has the same effect as using a pronoun in English. If this is really the same phenomenon as what we find in Spanish and Italian, then we would expect that Mohawk verbs also change their form. They should, in fact, go a step beyond Italian and Spanish, changing not only with the intended subject but with the intended object as well. This is true. The following examples show how the verb changes to express different subjects when the understood object is kept constant as 'it':

*k*enuhwe's	'I like it'
*s*enuhwe's	'You like it'
*r*anuhwe's	'He likes it'
*y*enuhwe's	'She likes it'
*yakwa*nuhwe's	'We like it'

These examples all have in common the basic verb root *nuhwe'*, meaning 'like,' and the ending *–s*, which expresses (roughly) the present tense. What changes is a prefix attached to the beginning of the word. Mohawk has even more prefixes than these to express other types of subjects, but these forms are enough to illustrate that Mohawk is like Spanish and Italian in this respect, making it no surprise that the subject can be omitted.

Now suppose we hold the subject constant—say, as the masculine singular 'he' form—and vary the intended object. Again, the prefix on the verb *nuhwe's* changes:

*Rake*nuhwe's	'He likes me'
*Ya*nuhwe's	'He likes you'
*Ro*nuhwe's	'He likes him'
*Shako*nuhwe's	'He likes her'
*Shukwa*nuhwe's	'He likes us'

In general, in order to decide what prefix to put on the Mohawk verb, one must know not only the intended subject but also the intended object, since the prefix depends on both. As a result, Mohawk speakers must learn many different prefixes: one for each possible combination of subject pronoun and object pronoun. Some simplifications apply, so there are "only" fifty-eight different prefixes rather than some 400, but this is still significantly more than the six different endings found in the Romance languages. This elaborate verb agreement system contributes to the impression that words in Mohawk are very complex. It also contributes to the fluidity of the syntax, helping to explain why both object and subject pronouns are redundant and therefore omittable in Mohawk. A one-word sentence like *Ranuhwe's* is not "missing" a subject or object after all; they are expressed inside the verb. The same completeness condition applies in Mohawk as in English, except that in Mohawk it requires the verb to have prefixes, whereas in English it requires the sentence to have pronouns.

This is not yet the whole story. I also need to explain why when a subject or an object noun phrase is included, it can appear almost anywhere in a Mohawk sentence. In Chapter 2 we saw that a noun phrase understood as the subject can sometimes follow the verb even in French and English, as long as a placeholder pronoun appears in the usual subject position:

> Il est arrivé trois hommes.
> It is arrived three men
> 'There have arrived three men.'

We took the inverted order in Italian (*Verrà Gianni* 'will-come Gianni') to be the same phenomenon, except that the placeholder pronoun is omitted, as redundant pronouns generally are in Italian. Perhaps something like this line of reasoning can be applied to account for the greater freedom of word order in Mohawk as well.

The key to accounting for this is recognizing that English and the Romance languages have sentences that involve what linguists call "left dislocation." An example is:

> That dress, Sak really likes it.

This sentence has a topic phrase *that dress* loosely attached at the beginning, before the subject. This topic phrase is associated with a matching pronoun that is in the normal direct object position (next to the verb, inside the verb phrase). English also allows right dislocation, in which the topic phrase appears at the end of the sentence:

> Sak really likes it, that dress.

The subject can also be dislocated, either to the left or to the right, so long as a pronoun is left in its place:

> Sak, he really likes that dress.
> He really likes that dress, Sak.

The ability of dislocated topic phrases to appear on either side of the sentence is a special case of a more general pattern; many "extra" phrases that are not part of the core sentence structure can be at-

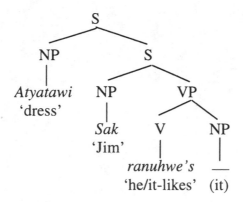

FIGURE 4.1   A Dislocation Sentence in Mohawk

tached in different positions. Adverbs like *yesterday*, for example, have a similar freedom: *Yesterday I bought a new car* and *I bought a new car yesterday* are equivalent variants of the same basic sentence. Sentences with dislocation are not all that common in English; they have a colloquial flavor and are found in informal speech more than in writing. They are also more common in the Romance languages than in English. But such constructions are possible to some degree in most languages, perhaps all.

Now imagine that Mohawk had a dislocation structure just like *That dress, Sak really likes it*. What would such a structure look like, given what we already know about the language? First, a noun phrase would appear at the beginning of the sentence, before the visible subject (if there was one). Second, there would be a pronoun object that matches the initial noun phrase in the usual object position. The only difference is that in Mohawk the object pronoun is not a freestanding word but rather a prefix on the verb. In the tree structure format introduced in Chapter 3, this sentence could be represented as in Figure 4.1. Since we do not see an independent object pronoun in Mohawk but only the "topic" noun phrase that it is associated with, it looks like the object comes before the subject—an apparent violation of the verb-object constraint. But this is an illusion; the real object is the prefix on the verb. This explains why "ob-

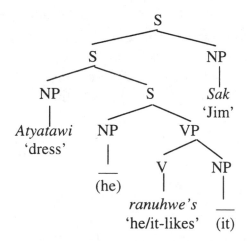

FIGURE 4.2 Object-Verb-Subject Order by Dislocation

ject"-subject-verb word order is possible in Mohawk in a way that is consistent with the verb-object constraint.

We can use this kind of reasoning to explain all the other word orders in Mohawk as well. We saw that in English either the subject or the object noun phrase can be dislocated. Furthermore, the dislocated noun phrase can appear on either the right side of the clause or on the left side. Finally, both subject and object pronouns are expressed as verbal prefixes in Mohawk, so no separate pronoun is ever needed. Given all this, it is easy to see that any word order can be derived as some combination of dislocating the subject and dislocating the object. For example, one of the rarest word orders is object-verb-subject order (see Chapter 3). But this order is reasonably common in Mohawk. It can be formed by dislocating the object to the left, while at the same dislocating the subject to the right, as in Figure 4.2. (Compare the English *That dress, he really likes it, Jim,* with overt subject and object pronouns). Any other order of verb, subject, and object is also derivable in this way, as interested readers can check for themselves.

In summary, Mohawk's nonconfigurational syntax emerges naturally as a byproduct of its being polysynthetic. Part of its polysynthesis is that its verbs inflect to show both their subjects and their objects. These inflections can count as subject and object pronouns, meaning that no other subject or object noun phrase is needed to make a semantically complete clause. If a Mohawk speaker chooses to make her sentence more descriptive by including a noun phrase anyway, that noun phrase will be in a dislocated position, linked indirectly to the subject or object prefix. As such, the noun phrase can be added anywhere, on the right or the left. This gives the impression that the language does not care about word order.

An important test of any theory is how it does on the details. Usually a variety of theories can account for the main effects. The correct explanation can often be distinguished from its competitors because it extends almost automatically to cover details as well, even details that were not noticed at first. So far I have argued that the main effects of word order in Mohawk can be explained by saying that Mohawk sentences are like dislocation sentences in English, with the prefixes on the verb counting as pronouns. Word order in Mohawk could also be explained in other ways, some of which may already have occurred to the reader. The dislocation hypothesis, however, has the great virtue of explaining various other quirks of Mohawk grammar. These additional consequences may not seem like big deals by themselves, but their cumulative effect can be substantial. (Readers who are not curious about these matters of detail are welcome to skip this section.)

One of these "minor" peculiarities concerns how Mohawk expresses so-called reflexive sentences. Reflexive sentences are a bit like those movies in which some talented comedian plays more than one role. They are sentences in which (for example) the doer of the action and the undergoer happen to be the same person. In English when one wants to refer to the same person more than once in a sentence, one normally uses an appropriate pronoun the second time:

John persuaded Sue that Mary likes him [*him* refers to John].

This general rule does not apply between the subject and the object of a simple sentence, however. In the following sentence, the pronoun *him* must be understood as referring to someone other than John:

John likes him [*him* does not refer to John].

To express the intended meaning, a special reflexive form of the pronoun must be used, one that includes the root *self*:

John likes himself.

Now let's compare this to Mohawk. Mohawk is like English in that the pronoun object of an ordinary verb cannot be interpreted as referring to the same person as the subject, even if it matches the subject in gender and number:

Sak   *r*onuhwe's.
'Sak likes him' [the liked person cannot be Sak].

Notice that the object pronoun is encoded as the verb prefix *ro-*, as usual in Mohawk. In spite of this superficial difference, the interpretation of the sentence is comparable to the English equivalent. Unlike English, however, this sentence cannot be made into a reflexive sentence by adding a noun phrase that has a special designated root meaning 'self.' Examples like the following are ungrammatical:

*Sak   ronuhwe's   rauha.
'Sak   likes   himself.'

(*Rauha* is a Mohawk word that can be translated as 'himself' in other contexts; it would show up in the Mohawk equivalent of 'Sak did it himself.') Reflexive sentences in Mohawk can be made only by adding a different kind of prefix to the verb, the so-called reflexive prefix *ratate-*.

Sak *ratate*nuhwe's.
'Sak likes himself.'

A quasi-English sentence that gives the flavor of the Mohawk form would be something like *Sak self-likes* or the more famous example *This tape will self-destruct in five seconds.* This, then, is a minor but intriguing difference between the grammar of English and the grammar of Mohawk. It is yet another instance of Mohawk's polysynthetic nature, in which grammatical relationships are expressed by building complex words rather than complex sentences.

This difference is in fact predicted by the analysis I sketched in the previous section. The basic idea was that the prefixes on the Mohawk verb are the equivalent of pronouns in English. These pronouns are present in every sentence; one does not have the option of simply leaving them off. You cannot say something like:

*Sak	nuhwe's	ne	atya'tawi.
'Sak	likes	the	dress.'

The verb has to be *ranuhwe's,* with the 'he + it' prefix *ra-.* Since subject and object pronouns are always present on the verb in Mohawk, other noun phrases must always be dislocated. For normal noun phrases, that is no problem. But reflexive noun phrases cannot be dislocated, even in English, as the following example shows:

   *John really likes him, himself.

(To the extent that this sentence is possible at all, it has another meaning, in which *himself* is an adverb associated with the subject *John* rather than with the object; it is equivalent to *John himself really likes him.*) This example contains a kind of internal contradiction. The reflexive form *himself* must refer to the same person as the subject *John,* that being its basic purpose in life. The reflexive form *himself* must also refer to the same person as the object pronoun *him,* since that pronoun occupies the position that *himself* was dislocated out of. An object pronoun cannot, however, refer to the same person as the subject of the clause, as we saw above. Sentences with a dislocated reflexive pronoun thus present an inescapable conflict. Now, in Mohawk all noun phrases must be dislocated. Therefore, it cannot have any reflexive noun phrases. It must find another way to express reflexive sentences, and special prefixes like *ratate-* do the job.

Other kinds of noun phrases that cannot be dislocated in English include nonreferential quantifiers like *everyone* and *nobody*. In simple sentences these expressions seem to act like ordinary noun phrases. The following examples, for instance, look parallel:

Chris walks much faster than I do.
Nobody walks much faster than I do.

Nobody is different from most other noun phrases, however, in that it doesn't refer to anything. For this reason, it cannot be the antecedent for a pronoun in a subsequent sentence. Lewis Carroll played with this property of *nobody* in the following passage from *Through the Looking-Glass:*

"Who did you pass on the road?" the King went on . . . .
   "Nobody," said the Messenger.
   "Quite right," said the King: "this young lady [Alice] saw him too. So of course Nobody walks slower than you."
   "I do my best," the Messenger said in a sulky tone. "I'm sure nobody walks much faster than I do!"
   "He can't do that," said the King, "or else he'd have been here first."

The basis of the joke is that the Messenger understands *nobody* in the usual way, as a nonreferential quantifier, whereas the White King is using it as an ordinary name for a person. Hence, the White King can replace *nobody* with the pronouns *him* and *he,* but the resulting sentences make no sense to the Messenger.

Since *nobody* cannot be the antecedent for a pronoun, and pronouns are crucially involved in dislocation structures, it stands to reason that *nobody* cannot be dislocated. This is true in English and other European languages:

> Chris, I saw her in the market yesterday.
> *Nobody, I saw her in the market yesterday.

Our hypothesis is that all independent noun phrases are dislocated in Mohawk. Therefore, we predict that Mohawk cannot have any non-referential quantifiers comparable to *nobody* in English. This is correct; examples like the following are systematically absent from Mohawk.

> *Sak   teshakokv   yah-uhka.
> Sak   he/her-saw   no + body

The negative element *yah* must be attached to the verb, not the noun, to create the Mohawk equivalent of a sentence like 'Sak didn't see anybody.'

The third "minor" difference between Mohawk and English is more subtle. Again it has to do with the conditions under which nouns and pronouns can be used to refer to the same thing. Consider the following sentence:

> John broke his knife.

This can easily be understood as saying that the owner of the broken knife and the one responsible for its breaking are the same person (John). But if the pronoun and the name switch places, so that the pronoun is the subject and the name is the possessor of the object, this interpretation is not possible:

> He broke John's knife.

In most contexts this sentence would be understood as meaning that some other person, different from John, broke John's knife. This suggests that some condition like the following is at work (the exact details are complex but not important to us).

THE REFERENCE CONDITION

Elements inside the object can depend on the subject for their reference, but not vice versa.

At first glance the negative part of this condition seems not to hold in Mohawk. In the following sentence the name inside the understood object noun phrase *can* refer to the same person as the pronominal subject:

Wa'thaya'ke'   ne   thikv   Sak   raoshare'.
He-broke-it   the   that   Sak   his-knife
'He broke that knife of Sak's.'   [The breaker can be Sak.]

But in fact there is no contradiction. The expression *ne thikv Sak raoshare'* is not technically the direct object in this sentence; rather, it is a dislocated "topic phrase." The true object is the pronoun 'it,' encoded in the verbal prefix. The reference condition doesn't rule out the possibility of a pronominal subject depending on something in a topic phrase for its reference. Indeed, this is possible in English:

John's knife, he finally broke it.   [*He* can refer to John.]

The same condition on pronoun interpretation can thus be applied to both languages. The results appear different because of differences in what counts as the subject and the object.

Overall, then, the hypothesis that all noun phrases in Mohawk are grammatically like dislocated noun phrases in English explains a number of subtle distinctions between the two languages that would otherwise be mysterious, in addition to the "first-order" effect that noun phrases are optional and freely ordered in Mohawk. And that is exactly how one recognizes any good theory.

———————

Let us put together the pieces of this story, so as to pinpoint exactly what underlies these various differences between Mohawk and English. In this discussion I have appealed to at least eight important grammatical similarities between Mohawk and English. They can be summarized as follows:

- Agents are subjects, and undergoers are objects.
- Objects combine with verbs before subjects (the verb-object condition).
- All the core participants of the action denoted by the verb must be expressed grammatically (the completeness condition).
- Pronouns can be omitted when their content is expressed by the verb.
- Dislocated noun phrases can be attached to either side of a clause.
- The pronoun object of a verb cannot refer to the same thing as the subject of that verb.
- Nonreferential quantifiers (e.g., *nobody*) cannot be dislocated.
- Elements inside the object can depend on the subject for their reference, but not vice versa (the reference condition).

These similarities show that, from a recipe perspective, Mohawk and English are not such different languages after all. The same principles and constraints operate in both.

What are the differences between Mohawk and English, from a recipe perspective? We have observed only two, and both were differences of degree rather than kind. First, Mohawk allows compounding in environments that English does not. It allows a noun to compound with a verb as a way of expressing the direct object of that verb (noun incorporation). It also allows two verbal elements to combine, in the special case where one verb counts as the object of the other (causative formations). Second, Mohawk has inflections that go on verbs that agree not only with the subject (as in Spanish) but also with the object. These properties indirectly cause many other differences between the two languages, but the root differences are simple.

Moreover, there is reason to think that these are really only one difference in two guises. Incorporation and agreement prefixes are both ways of expressing the object of a transitive verb. Both make it part of the complex verb. We can say the core difference that underlies all the others is that Mohawk sentences must satisfy one extra condition:

THE POLYSYNTHESIS PARAMETER

Verbs must include some expression of each of the main participants in the event described by the verb (the subject, object, and indirect object).

This polysynthesis parameter can be thought of as a variant of the completeness condition. It not only says that all participants must be expressed but puts a language-specific condition on how they are to be expressed. This condition clearly does not apply to English. Thus, it is a parameter—a recipe statement that is part of some (I-) languages but not others. The noun-verb compounding found in Mohawk is one way of honoring this condition because the incorporated noun expresses the object that undergoes the action inside the verb. English imposes no such requirement; therefore it does not bother with noun-verb compounding. Having a prefix on the verb that counts as an object pronoun is another way to honor the condition. Spanish, which has no such condition, does not bother with verb-object agreement.

Confirmation for this unified way of looking at Mohawk comes from an interesting pattern of facts that involves both object prefixes and noun incorporation:

*Sak	ra-nuhwe's		ne	owira'a.
Sak	he-likes		the	baby

Sak	shako-nuhwe's		ne	owira'a.
Sak	he/her-likes		the	baby

Sak	ra-wir-a-nuhwe's.
Sak	he-baby-likes

*Sak	shako-wir-a-nuhwe's.
Sak	he/her-baby-likes

All four versions of this sentence attempt to say approximately the same thing: that Sak likes the baby. The first version has neither noun incorporation nor agreement with the feminine object. (The prefix *ra-* shows only that the subject is third-person masculine; it would be appropriate if the verb were intransitive, but not otherwise, except as below.) This sentence clearly violates the polysynthesis parameter, because there is no indication of the liked one on the verb. The second version of the sentence does have the proper prefix for showing that there is a third-person feminine object, so it is grammatical. The third version does not indicate the object by verb agreement; it has the same "subject-only" prefix, *ra-*, as the first sentence. Nevertheless, it is grammatical because the object noun *wir* 'baby' has been incorporated into the verb, thereby expressing the object of the verb by the second technique. Finally, the fourth sentence shows that it is bad to have *both* an incorporated noun *and* an agreement prefix to express the same object. One or the other is needed to fulfill the polysynthesis parameter, but having both is superfluous and generally avoided. This complementarity between incorporation and agreement shows that the two work together as part of the same system. They are not merely two independent peculiarities of Mohawk grammar.

The polysynthesis parameter also requires that Mohawk verbs must include exactly one expression of the subject of the verb. The choices are more limited in this domain. The technique of noun-verb compounding cannot express the subject of the verb because of the verb-object constraint. Using agreement affixes on the verb is therefore the only possibility. In using such affixes, Mohawk is not strikingly different from Spanish and Italian—yet even here we can detect a small distinction. Although it is true that all *tensed* verbs in Spanish agree with their subjects, Spanish also has infinitival verbs. Infinitives are "bare" verb forms that do not have tense marking or agreement with a subject. For instance, in the following example, the main verb *quiero* 'want' bears the characteristic -o ending for a first-person singular subject, but the infinitival verb *comer* 'eat' does not, even though 'I' is understood as the subject of both verbs:

[Yo]	quiero	comer	las manzanas	amarillas.
I	want	to-eat	the apples	yellow

'I want to eat the yellow apples.'

Mohawk has no comparable infinitival verb forms. In the Mohawk equivalents of Spanish sentences with infinitives, both verbs must bear the same subject agreement (here *yukwa-/yakwa-*, indicating a first-person plural subject 'we').

Te-*yukwa*-tuhwvtsoni	a-*yakwa*-hsere'	ne	kayanere'kowa.
We-want	we-follow	the	great-good

'We want to follow the League of Peace.'

Mohawk is thus more serious about having an expression of the subject on every verb than Spanish is. In European languages some verbs include expressions of some participants, but in Mohawk this is systematic and obligatory. In other words, Mohawk obeys the polysynthesis parameter and European languages do not.

It must be emphasized that this is the *only* significant difference we have discovered in the recipes for the two kinds of languages. Whereas Mohawk and the European languages are alike in (at least) eight substantive ways, their differences can all be traced back to one fundamental difference in how grammatical relationships are expressed. Once again, we see that how different two languages are depends on exactly what we mean by "language." If one compares Mohawk and English sentence by sentence, E-language-style, they are completely different. Zero percent of actual Mohawk sentences have the same structure as their English counterparts. This is because the polysynthesis parameter affects the shape of every single sentence of Mohawk. But if one thinks of languages as I-languages—as sets of principles and constraints that determine the structure of sentences— then Mohawk and English turn out to be very similar. They have (at least) eight principles in common and only one that is different, making them roughly 89 percent the same in terms of their recipes.

In my book *The Polysynthesis Parameter,* I argued that many more principles can be found that are valid for both Mohawk-type languages and English-type languages but maintained that the major differences can be traced back to the polysynthesis parameter. In this respect, the results of modern linguistics are much like the results of modern biology, which has shown that two mammals—say, a mouse and an elephant—may look quite different, but their underlying genetic codes and biochemistry are almost identical. Just as the recipe for a mouse is remarkably similar to the recipe for an elephant, so the recipe for Mohawk is remarkably similar to the recipe for English. You just add the equivalent of a tablespoon of yeast when making Mohawk. The parametric hypothesis is powerful enough to handle even the largest-looking differences among languages with only a small number of parameters.

———

Before leaving this chapter, let us briefly consider a more global perspective on these issues. We already know from Chapter 3 that many widely scattered languages—such as Edo and Thai and Zapotec— have the same gross syntactic system as English. Are there also other languages like Mohawk? The answer is yes. There are not a huge number of them, and most do not have many speakers. You are unlikely, while riding the bus to work, to overhear someone speaking to a friend in a full-fledged polysynthetic language. Nevertheless, there

TABLE 4.1   Polysynthetic Languages of the World

Language Family	Sample Languages	Where Spoken
Caddoan languages	Wichita	American Great Plains
Tanoan languages	Southern Tiwa, Jemez	New Mexico
Nahuatlan languages	Nahuatl (esp. Classical)	Central Mexico
Gunwinjguan languages	Mayali, Nunggubuyu, etc.	North central Australia
Paleosiberian languages	Chukchee, Koryak	Northeastern Siberia
Mapuche	Mapuche	Central Chile
Ainu	Ainu	Northern Japan
Munda languages?	So:ra?	India

are languages that are historically unrelated to Mohawk but that have very similar grammars in the respects we have reviewed. These are listed in Table 4.1. And I would not be surprised if a few more polysynthetic languages came to light, particularly in the less-studied areas of South America or New Guinea.

That Mohawk is not unique gives indirect support to the polysynthesis parameter. If one did not believe in parameters, one would presumably say that languages acquire their distinctive properties by a gradual process of cultural evolution. In this way many small changes could accumulate until eventually a very different language was created. At first glance this picture seems highly plausible, and some aspects of language do evolve in just this way. For example, the vocabulary of a language changes gradually as words come into the language, mutate, and die out over time. But if all aspects of a language developed like this, then we would expect each language to be a unique reflection of the cultural history of its people-group. It would be a far-fetched coincidence if two unrelated languages, spoken on opposite sides of the world, ended up being essentially the same by this kind of gradual and accidental process. It would be the equivalent of two people's independently tossing a coin 100 times and getting the same sequence of heads and tails—a practical impossibility. Indeed, the vocabularies of historically unrelated languages are never substantially the same: Your Mohawk-English dictionary will never do double duty, for example, as a Mayali-English dictionary.

But in the domain of grammar we find that languages are not so unique. Although the total number of imaginable grammatical systems is in theory quite large, in practice we find relatively few systems being used over and over again in different parts of the world. Your Mohawk grammar could serve as a decent Mayali grammar, to a first order of approximation. This strongly suggests that the distinct grammatical systems we see in the world did not evolve by a gradual and unconstrained process of cultural evolution. It is, however, what we expect to find if Mohawk is the product of adding a single ingredient—the polysynthesis parameter—to the generic recipe for human language. On this view, languages face essentially a two-way choice, whether to include the polysynthesis parameter or not. It is not at all surprising that unrelated languages happen to make the same choice and end up being grammatically similar. This is the equivalent of having two people independently tossing a coin once and both having it come up heads: One expects it to happen quite often.

The Gunwinjguan languages of Australia are thought to be historically related to the other languages of Australia—not surprising, given Australia's geography. Many Australian languages have few or none of the distinctive polysynthetic properties seen in Mohawk: They do not have noun incorporation, for example, or object agreement prefixes or subject agreement prefixes. Yet the Gunwinjguan languages have developed all of these features. This gives further credence to the idea that these are not unrelated developments but the varied ramifications of one unified development.

Table 4.1 also tells us something worth knowing about the relationship between language and the physical and cultural environment of the people who speak it. It tells us that they are independent of each other. One might naively have thought that different languages would be adaptations to different physical environments. One of the remarkable features about human beings from a biological point of view is our ability to prosper in almost any ecological situation. Perhaps our capacity to use different kinds of languages was an important factor in achieving this kind of flexibility. Arguably the most famous myth about language is that the Eskimos have more than thirty different words for snow. This sounds like a clear case of the adaptation of language to distinctive properties of the environment.

The myth is not true, however. The Eskimos in fact have fewer words for snow than the average skier. Anthropologist Laura Martin and linguist Geoffrey Pullum have documented how this myth arose, beginning with a technical remark by Franz Boas in 1911. This remark was misremembered in a famous 1940 article by an amateur linguist, Benjamin Whorf, already committed to the idea that speakers of different languages see the world differently. From there it was picked up by popular sources, who further distorted and magnified the numbers with each retelling, until the falsehood was firmly embedded in the lexicon of American folk beliefs.

When one turns from superficial features of the vocabulary to basic grammatical systems, one finds even less reason to believe the adaptationist view. Polysynthetic languages are spoken in almost every environment humans inhabit, ranging from arctic tundra (Chuckee) to temperate forests (Mohawk) to arid plateaus (Nahuatl) to tropical brushlands (Mayali). Languages that are not polysynthetic are also spoken in all these different ecosystems. Apparently, then, language and physical surroundings are mutually independent. Modern population migrations tell us the same thing: English, Spanish, and Chinese are now successfully spoken all over the world without changing grammatically. No doubt the capacity for language has contributed to the human ability to adapt to so many different physical situations, but the fact that language includes parameters that allow for syntactic variation does not seem to be a factor in this.

Similar reasoning also suggests that language is largely independent of other features of culture—although this is harder to pin down, simply because it is harder to be precise about what culture is. Benjamin Whorf, Edward Sapir, and other anthropological linguists from the first half of the twentieth century popularized the view that the language people speak influences their basic worldview and perceptions in important ways. This intriguing idea is one of the things that attracts people to linguistics even today, and it has lately enjoyed something of a renaissance among specialists. The more one knows about the details of how languages vary, however, the less likely it seems that this idea can be spelled out in a substantive and interesting way.

Suppose I am correct in saying that the primary difference between a Mohawk-style language and an English-style language is that verbs in Mohawk must contain an expression of each participant in the action denoted by the verb. What causal interaction could this possibly have with other features of traditional Mohawk culture? The answer obviously depends on the tricky issue of just what one takes traditional Mohawk culture to be. But there is no discernible connection between this grammatical feature and the kinds of cultural facts that are the staples of anthropological description. In traditional Mohawk communities, kinship was matrilineal, tobacco was burned in religious ceremonies, people lived in longhouses, they practiced slash-and-burn agriculture, and they engaged in ritual torture of their enemies. What could any of this have to do with noun incorporation or object agreement? The polysynthesis parameter seems relevant only to the domain of language. Nor do the linguistic differences give us any reason to suspect that Mohawk speakers think differently from English speakers in any fundamental sense. On the contrary, the linguistic representations that the Mohawks entertain can be matched with English ones in a complex but algorithmic way, and the principles of deduction that relate linguistic recipes to actual sentences must be essentially the same. Mohawk and English are different but commensurable, and the differences seem keyed specifically to language. In the most general sense, it is a tautology that language is an important part of the culture of a people. But particular linguistic differences do not seem to be tied to other particular cultural differences in any mutually illuminating way.

Table 4.1 does suggest one property that all cultures associated with robustly polysynthetic languages have in common. The suggestion is just true enough to be dangerous. The polysynthetic languages are all "local" languages, spoken in relatively small traditional communities.

None of them is the official language of a modern, technologically advanced nation-state. It has been suggested that this not an accident, that a polysynthetic language could arise and flourish only in small, face-to-face communities where one tends to interact with the same people throughout one's life. A polysynthetic language, it is said, could not be the language of mass literacy and an impersonal bureaucracy.

This claim seems dubious to me—indeed, it borders on racism. No precise reason has been proposed for why polysynthetic languages would work only in a certain kind of society. Nor does this suggestion account for the actual distribution of languages in the world, except in the crudest way. For example, one of the most technologically advanced, socially stratified, and imperialist cultures of pre-Columbian America was that of the Aztecs. It has been estimated that they were equal to their Spanish conquerors in many of these respects. Nevertheless, the Aztecs' language (Nahuatl) was one of the most polysynthetic languages of the Americas. Nor are the speakers of Australian polysynthetic languages significantly different in their material culture or social organization from their neighbors who speak nonpolysynthetic languages. Thus, there is little evidence of a correlation between polysynthesis and type of society. Jerrold Sadock informs me that the introduction of a nation-state bureaucracy into Greenland has tended to increase rather than decrease Greenlanders' use of polysynthesis, simply because it creates an impetus for making up difficult-to-understand new technical terms, as bureaucracy does everywhere. For Greenlandic Eskimo, the natural way to meet this demand is by using the polysynthetic resources that are available in that language.

The tendency of modern bureaucratized nation-states to speak nonpolysynthetic languages can be easily explained as a historical coincidence. Modern nation-states developed only in the past few centuries and primarily in Europe, a small continent where no polysynthetic languages were spoken. If geohistorical factors had conspired to give the Aztecs and the Mohawks most of the guns, germs, and steel when the continents came into contact, we would be discussing the opposite hypothesis now. You would be reading a Nahuatl book that refuted (I hope) the idea that head-initial languages with simple words are spoken only by small, primitive cultures. The world is a small enough place that historical accidents can affect its overall look.

The unity of the polysynthesis parameter should show up not only in the distribution of languages throughout the world and in the way languages change toward stable patterns but also in the way children learn their first language. One aspect of the Code Talker paradox is that someone as seemingly unsophisticated as a human child can learn something as complex as a natural language quickly and accurately. The parametric theory helps to explain how this is possible. Since most of the recipe for speaking Mohawk is the same as the recipe for speaking English or Japanese, most of it could be encoded in the child's mind from the start. It would therefore not have to be learned from direct evidence, in the usual sense of learning. Principles like the eight I gave above are equally applicable to the understanding of any human language. On this view, children are rather like computers that come from the factory with much of their software preloaded. They do not have to program an entire language from scratch; they only have to set a few parameters—the linguistic equivalent of clicking on a few items from the "Options" menu—and they are ready to go.

We would predict, then, that children should catch on to the various aspects of polysynthesis more or less simultaneously. They should not have to figure out separately that a language has noun incorporation, that it has object agreement, that it has no nonreferential quantifiers, and so on. Any one of these facts could serve as a clue to young children that the language being spoken around them is a polysynthetic one. Once they have established that, they can deduce the language's other characteristic properties for themselves. As a result, they should acquire the different features of polysynthesis more or less all at once.

Predictions of this type are known to hold in other cases. For example, children do seem to catch on all at once to the fact that heads come at the end of phrases in languages like Japanese and Turkish. They are not distracted by the logical possibility that some phrases might be head-initial and others head-final. But practical problems have made it hard to determine whether they also catch on to polysynthesis all at once or not. The world is a small place, and most polysynthetic languages are spoken only by a few small and shrinking communities of speakers. It is usually impossible to find a critical mass of children learning these languages at the university daycare, ready for experimentation. This creates an unfortunate gap in the evidence for the polysynthesis parameter.

It is perfectly imaginable that *all* of the polysynthetic languages will become extinct in the next fifty years or so. If so, the world will be an even smaller and duller place as a result. We will know that speaking this intriguing kind of language is a possible way of being human only in the way that we know that a passenger pigeon is a possible way of being an animal—only by rumor and dusty records.

5

Alloys and Compounds

S OME PHYSICAL SUBSTANCES ARE MADE UP of only one kind of atom. Argon gas is made up of single argon atoms; oxygen gas is a molecule made up of two oxygen atoms; graphite and diamonds are carbon crystals. If that were all there was to chemistry, there would be only about 100 distinct physical substances in the universe. But a knowledge of the various elements is not the end of chemistry; rather, it is the beginning. Chemistry studies not only the 100-odd elements in isolation but also how those elements combine into a myriad of mixtures, alloys, and compounds whose properties are complex functions of the properties of their atomic parts. Chemistry also explains why some combinations of atoms are impossible as compounds—or highly unstable, so that they exist only briefly under special conditions. For example, it follows from the theory of valence that hydrogen and oxygen can combine in ratios of two to one to make water, or two to two to make hydrogen peroxide but not in a ratio of one to two or two to three.

Similar issues arise with linguistic atoms. So far, I have concentrated on clear, paradigmatic examples of parameters. I have compared languages that are "minimal pairs" in the sense that their differences can be attributed to a difference in a single parameter, at least over the range of data considered. Thus, Japanese word order differs from English only in that phrases have the head at the end rather than at the beginning, and Mohawk sentence structure differs from English only in that all participants have to be expressed within the verb. Moreover, these properties are expressed uniformly throughout the entire language. Every kind of phrase is head-final in Japanese, and every kind of participant must be represented in a complex word in Mohawk. This is the linguistic equivalent of studying pure argon and comparing it to pure oxygen or pure carbon. Valuable as it is for isolating some basic parameters, it is not the full story for linguistics any more than it is for chemistry. If it were, then there would be only a small number of language types, with sharply different properties.

Languages can differ in more than one parameter. When this happens, those parameters will interact in ways that establish the basic properties of these "alloy" and "compound" languages. Ideally, these interactions should be deducible from the logic of the individual parameters, just as properties of chemical compounds are deducible in part from the properties of the atoms that make them up.

In this chapter I discuss some extensions of the parametric theory along these lines and how they might account for various blends and minor variations on the main language types presented so far. We will also see how the parametric theory can explain why some of these "blended types" are quite common, whereas others either do not exist at all or else are rare and unstable. This topic is at the frontier of linguistic research, and there is little consensus about precisely how some of the particular illustrations I present should be analyzed. It is becoming clear, however, that the conceptual tools for explaining them lie within the parametric theory.

———————

Back in Chapter 3 I emphasized the remarkable degree to which the basic order of words in English is the same as the basic order of words in languages like Edo, Indonesian, and Zapotec. Similarly, the basic order of words in Japanese, although very different from the

English order, is similar to the order of words in languages like Lakhota, Turkish, and Quechua. Yet striking as this is, it would be an exaggeration to say that English and Edo or Japanese and Quechua are identical in their grammars (meaning that they are the same except for their repertoires of basic words and sounds). Also, at the end of the previous chapter I claimed that there are other polysynthetic languages that are very much like Mohawk, including Nahuatl, Mayali, and Chukchee. But again none of these languages is grammatically *identical* to Mohawk. These "types" are not rigid templates; they are more like themes, upon which natural language allows certain variations.

Since the parameter is our tool for explaining differences among languages, we want to interpret these variations as involving parameters as well, but on a different scale. For example, English and Edo share the same setting for the head directionality parameter (the head is initial), the subject side parameter (subjects are initial), the polysynthesis parameter (it doesn't hold), and the null subject parameter (sentences require subjects). Yet they could still differ in their settings for one or more additional parameters that happen to make only a relatively small difference in the qualities of the E-language as a whole. In terms of my overarching chemical analogy, English and Edo could be thought of as two alloys of iron, differing from each other roughly as stainless steel differs from black steel. There is no such thing as a steel atom; steel is the result of mixing iron atoms with a small percentage of carbon. Different alloys of the same metal are typically more similar to each other than different elements are; steel is more like cast iron, for example, than it is like chlorine gas. Nevertheless, different alloys have significant "second-order" differences in their properties, so that some alloys are better for one application and less good for others. In a similar way, Edo and English would have mostly the same parameters but with a small amount of parametric difference added in.

The idea that some parameters have more noticeable effects than others is not at all surprising or hard to understand. On the contrary, it is exactly what one should expect. A small difference can have large repercussions, but only when special conditions hold—only if it is perfectly situated so as to maximize its interactions with other elements of the system. In other circumstances a small change will have only a small effect. Adding a tablespoon of yeast to a recipe has a large impact on the final product, but a tablespoon of salt is not nearly as noticeable. From the point of view of recipe writing, there is essentially no difference between the two: The instruction to add a tablespoon of yeast is hardly different from the instruction to add a tablespoon of salt. The difference comes entirely from the way yeast interacts with other ingredients already present in the bread dough. Similarly, a loose rock can cause a landslide—but only if it is at the top of the slope.

The same is true for parameters. Some simple changes in the recipe of a language have repercussions that affect the overall "feel" of the language in striking ways. These are the kinds of parameters that I have focused on so far. Other changes have the same logical character, but because of the way they happen to interact with other features of language their effects don't spread through the language in the same way. These parameters do not create new types of languages but variations on one of the main types.

———————

A good illustration of how these notions work out in practice comes from comparing English with its neighbor language Welsh. Word order in Welsh is very much like word order in English, with one salient difference: Subjects come before the main verb in English, whereas they come after it in Welsh. This can be seen in each of the following sentences:

Darllenais   i   y   llyfr.
Read   I   the   book
'I read the book.'

Euthum   i   â   Mair   i'r   sinema.
Went   I   with   Mary   to-the   cinema
'I went with Mary to the movies.'

Disgwyliais	i	yr	ennillai	John.
Expected	I	that	would-win	John

'I expected that John would win.'

Apart from this peculiarity, Welsh looks like an ordinary head-initial language: Verbs come before objects, prepositional phrases, and embedded clauses; prepositions come before their noun phrases; articles come before nouns; and complementizers come before the clauses they introduce. English and Welsh clearly both have the head directionality parameter set in the same way: Words are added before phrases rather than after them. Since this parameter is relevant to the construction of every phrase in a sentence, it has massive superficial consequences. Nevertheless, there must also be at least one other parameter whose effects are relatively minor but that determines where the verb is placed with respect to the subject. Welsh must be an alloy of the canonical English type.

From a global perspective, the Welsh word order is not as common as the English order, but it is not especially rare either. Most counts put it at around 10 percent of the languages of the world, compared to 40–45 percent for the English order. This is enough for the Welsh order to qualify as the third most frequent word order type (after the English type and the Japanese type) based on Tomlin's sample of 402 languages selected from each language family (see Table 5.1; Mohawk-type languages are not included).

Most of this table we already understand to a fair degree, based on the discussion in Chapter 3. The first two lines give the two most common word order types; they are roughly equally frequent and together they account for almost 90 percent of the world's languages. The difference between these two language types is determined by the head directionality parameter. We also understand the bottom half of the table, for the most part. In Chapter 3 I entertained the possibility that some languages allow subjects to be added at the end of other phrases rather than at the beginning—the so-called subject side parameter. When this option is taken by a head-first language, the result is verb-object-subject word order, employed by languages

TABLE 5.1   Percentages of Languages with Various Word Orders

Basic Word Order	Percentage of Languages	Example Languages
Subject-[verb-object]	42	English, Edo, Indonesian
Subject-[object-verb]	45	Japanese, Turkish, Quechua
Verb-subject-object	9	Zapotec, Welsh, Niuean
[Verb-object]-subject	3	Tzotzil, Malagasy
[Object-verb]-subject	1	Hixkaryana
Object-subject-verb	0	(Warao)

like Tzotzil. When this option is taken by a head-last language, the result is a language with object-verb-subject order, like Hixkaryana. (Left unexplained was why so few languages, just 4 percent, use this option of putting subjects at the end. I return to this from a different perspective in the next chapter.) We even have an account of why object-subject-verb languages do not exist—although there are reports that one or two South American languages such as Warao have this word order. Such languages would violate the verb-object condition (stated in Chapter 4), which says that the verb and the object must form a phrase that does not include the subject. This verb phrase is enclosed in brackets for the other four language types in Table 5.1. No such phrase, however, could be constructed in an object-subject-verb language. This could explain why object-subject-verb languages are nonexistent or at least extremely rare.

In this otherwise rosy picture, verb-subject-object languages like Welsh stick out like a sore thumb. The kinds of principles and parameters that we have discussed so far have no way of forming such sentences, because they, too, violate the verb-object constraint. All things being equal, one would predict this type of language to be just as rare as the object-subject-verb type. But it is not. Although one might write off the few reports of object-subject-verb word order as minor anomalies, that is clearly impossible for the verb-subject-object order of Welsh. Although a minority language type, it is not an inconsequential minority. Such languages are more frequent than the

Tzotzil type and the Hixkaryana type, which fit easily into the parametric theory. Moreover, verb-subject-object languages are found in historically unrelated pockets around the world. In addition to the Celtic languages, this word order is found in some Afroasiatic languages (Berber, Arabic), in some Southeast Asian and Oceanic languages (Niuean, Chamorro), and in various American languages, including the Salish languages of the Pacific Northwest and many languages of Mesoamerica. Moreover, some of these languages are relatively accessible to scholars and thus have been studied in detail—especially Welsh, Irish, and Arabic. There is no chance that the reports of verb-subject-object order are mistaken or that this order is the result of unique historical factors.

Fortunately, we are in a position to show exactly how a verb-subject-object language can arise as a result of two additional, second-order parameters. (There may be more than one way this comes about, but we will be content to find one.) A hint as to what these additional parameters might be comes from Welsh sentences that contain a separate tense auxiliary as well as a main verb. Our first three examples of Welsh had no auxiliaries at all. Since Welsh is a head-first language, we expect an auxiliary to come before the main verb, and that is correct. What is interesting is that the subject comes between the auxiliary and the main verb:

Naeth	y	dyn	brynu	car.
Did	the	man	buy	car

'The man did buy a car.'

(This order is possible in English, too, but only as a question; in Welsh it counts as an ordinary declarative statement.) It is significant that we find English-like subject-verb-object word order in this particular situation, even in Welsh.

In Chapter 3 we saw that the position of subjects is not determined by the head directionality parameter. Rather, subjects are noun phrases that are merged together with some other phrase. As a result, the subject appears at the beginning of the clause in both

English-type and Japanese-type languages. We do not want to change this principle of subject placement very much for Welsh, since the Welsh subject still appears relatively early in the sentence, before the object, any prepositional phrases, or an embedded clause. It even comes before the verb when an auxiliary is present. What we can say is that there is a difference between Welsh and English in the phrase to which the subject attaches. In English the subject combines with the auxiliary phrase formed by merging an auxiliary with a verb phrase. This accounts for the placement of the English subject before the auxiliary as well as before the verb. The difference could be that in Welsh the subject merges with the verb phrase first, and then the auxiliary combines with the newly formed unit to make an auxiliary phrase. In tree notation this contrast between English and Welsh could be represented as in Figure 5.1. (Here *S* is a special notation for the phrase formed by joining a subject to any other phrase. This extra symbol is necessary because the phrases so formed do not have heads that they can be named after, according to the simple terminology of Chapter 3.) This creates the right word order for Welsh sentences that have an auxiliary. This Welsh structure does not violate any principles of language that we have stated so far. In particular, it respects the verb-object constraint because the object combines with the verb before the subject does. The Welsh structure also conforms to all the same word order parameters that the English one does. Thus, we are not actually changing anything; we are merely exploiting a loophole in the system. The difference between the two languages can be stated in parameter form as:

THE SUBJECT PLACEMENT PARAMETER

The subject of a clause is merged with the verb phrase (as in Welsh).

or

The subject of a clause is merged with the auxiliary phrase (as in English).

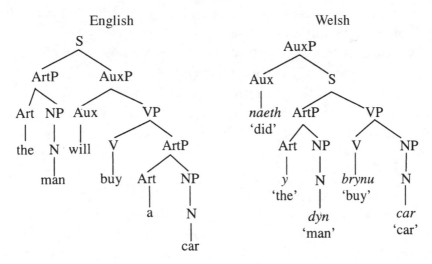

FIGURE 5.1   Subject Placement in English and Welsh

We can now come back to the verb-initial sentences   that sparked our original interest in Welsh—sentences like the following:

Bryn-odd	y	dyn	gar.
Buy-PAST	the	man	car

'The man bought a car.'

This sentence differs in two ways from the one we already analyzed. Not only is the position of the verb with respect to the subject different, but also the past tense and the verb are no longer expressed by two separate words. Rather, both elements of meaning are expressed by one complex word, consisting of a verb root and a suffix that marks the tense. In this respect verb-initial sentences in Welsh are a little bit like the polysynthetic sentences of Mohawk, in which a complex word takes the place of the multiword syntactic constructions found in other languages. To be precise, in Welsh it is not only the verb that appears before the subject, but the tense marker as well. For the verb, this presubject position is somewhat surprising, but for the tense marker it is not surprising at all. It is normal for the expression of tense to come before the subject in Welsh, given the subject placement parameter.

We can capitalize upon this as follows. Suppose that the tense marker and the verb start out as separate elements in this sentence, just as they did in the previous example. When the particular tense that the speaker has in mind happens to be an affix, the tense auxiliary needs to "fuse" with a nearby verb that can bear it. This can be thought of as the same process that fuses the object and the verb into a single word in Mohawk. That is all well and good, but now the verb and the tense marker in Welsh must face the difficult question that confronts lovers of all kinds: "My place or yours?" In grammatical terms, the question is whether the verb will be drawn to the original position of the tense auxiliary or whether the tense auxiliary will be drawn to the original position of the verb. It is reasonable to suppose that this indeterminacy is resolved by another parameter:

THE VERB ATTRACTION PARAMETER

Tense auxiliaries attract the verb to their position.
or
Verbs attract tense auxiliaries to their position.

Welsh is a language in which the tense auxiliary plays the host. It attracts the verb to itself, causing the tensed verb to be displaced out of its original (and expected) position after the subject. The verb ends up before the subject because that is where its host is. This can be represented graphically, as in Figure 5.2.

This, then, is a way of accounting for verb-initial orders in languages like Welsh. Those orders are a result of not one but two additional parameters: one involving the placement of the subject, the other involving the displacement of the verb. In their overall logic, these two new parameters are identical to all the others we have seen. They are simple recipe statements about how to go about constructing sentences. In fact, they complement the parameters we have already seen by resolving fine points those parameters leave open. But

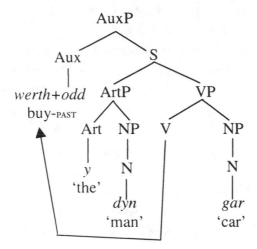

FIGURE 5.2   Verb-Subject-Object Order in Welsh

these two parameters are relatively narrow and specific: One has to do only with subjects; the other mentions verb and tense relationships by name. As a result, it is hardly surprising that the effects of these parameters are localized to a particular configuration. In this respect they are different from the head directionality parameter and the polysynthesis parameter, which shape every phrase of every clause. The subject placement parameter and the verb attraction parameter are like loose rocks near the bottom of a mountain slope: They cannot cause a landslide because they are not in a position to bump into many things.

———————

At first glance it might seem strange that the extensive differences between English and Japanese word order should all be attributable to a single parameter, whereas the small difference between English and Welsh requires two. Nevertheless, this somewhat surprising proposal makes sense of the relative frequencies of the word order types given in Table 5.1. As a simplifying idealization, suppose that each linguistic community of the world decided which parameters its language would include not by a long historical process (which I take up in

Chapter 7) but rather by the flip of a coin. Since the difference between English-style word order and Japanese-style word order is attributable to a single parameter, there is only one decision to make
by coin flip: heads, heads are initial; tails, heads are final. So we expect roughly equal numbers of English-type and Japanese-type languages. Table 5.1 confirms that this is so. Within the head-initial
languages, however, it requires two further decisions to get a verb-
initial, Welsh-type language: Subjects must be added early *and* tense
auxiliaries must host verbs. If either of these decisions is made in the
opposite way, then subject-verb-object order will still emerge. If the
decisions were made by coin flips, we would predict that about 25
percent of the head-initial languages would be of the Welsh type and
75 percent of the English type. This, too, is approximately correct:
By Tomlin's statistics, about 18 percent of head-initial languages are
Welsh-type, verb-subject-object languages (other counting schemes
put the figure somewhat higher). The match is not perfect, but it is
about as good as one could expect, given that the parameter settings
of a language are not really decided by coin flips. They are determined by a historical process that tends to maintain the parameter
settings of one's forebears, although with an element of chance due
to "noise" in the learning process. The world is small enough that a
few cases of (say) English speakers expanding at the expense of
Welsh and Irish speakers could easily skew the figures away from the
theoretical ideal. The crucial point is that a number of factors have
to fall together in just the right way for a Welsh-type language to appear, whereas it is relatively easy to get an English-type or Japanese-
type language. In this respect, our analysis has the right flavor.

    If it is true that Welsh and English differ in two parameters, then
(all things being equal) we would expect to find languages that are
halfway between them in terms of I-language. There should be languages that have the English setting of the subject placement parameter but the Welsh setting of the verb attraction parameter or vice
versa. Since all of these languages would look rather like English in
their gross word order, the differences will not show up in rough typological counts such as Tomlin's. Nevertheless, one might hope to

detect a difference by closer analysis. This also seems to be correct, at least to some extent. Let us consider languages that attach subjects to auxiliary phrases (like English) and imagine the effect of varying the setting of the verb attraction parameter. At first glance it might seem as if this would make no detectable difference. Tense auxiliaries and verbs in English-like languages are in the enviable position of lovers who also happen to be next-door neighbors; for them the question "My place or yours?" does not pose as much of a dilemma. But there might be linguistic elements other than the subject that can reveal how this question is answered. Adverbs, for example, are placed between the tense auxiliary and the verb in English and French:

Jean a souvent embrassé Marie.
John has often kissed Marie.

Apparently, adverbs attach to the left side of the verb phrase to form a larger verb phrase in both languages. (Most adverbs can also attach on the right side of the verb phrase, forming a sentence like *John has kissed Marie often;* see Chapter 4.) Such adverbs are in essentially the same position as subjects are in Welsh. But English and French have different word orders when a single main verb bears the tense inflection. In French the inflected verb comes before the adverb, whereas in English it comes after:

Jean *embrasse souvent* Marie.
John *often kisses* Marie.

(Readers interested in learning to feign a French accent should note that using sentences like 'John kisses often Mary' is an excellent supplement to saying *ze* and *zat* for *the* and *that,* since French speakers have a difficult time mastering this word order.) This fact is explained immediately if we say that French and English differ in their settings for the verb attraction parameter: Tenses host verbs in French, but verbs host tenses in English. The difference is illustrated in Figure 5.3. French is thus one of the intermediate languages we

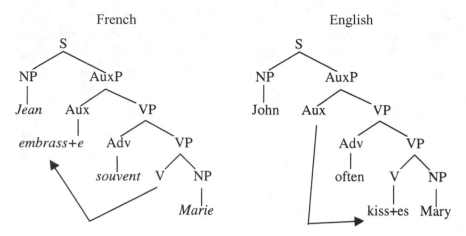

FIGURE 5.3   Verbs and Adverbs in French and English

were looking for: Its subject placement is like English, but its verb attraction is like Welsh. The two differences between English and
Welsh can in principle be dissociated, confirming that two separate
parameters are at work.

———————

So far I have shown that there is a variant of the English-type, head-
first language that is induced by two additional parameters. Now we
should go back and see how these new parameters affect the rest of
our word order typology. How, for example, do these new parameters affect Japanese-style, head-last languages? We might initially expect to find variations on the Japanese theme that parallel the
variations on English. But that is not what we observed in Table 5.1;
there is no fourth type of language that is generally like Japanese except for differences in the placement of the subject and the tensed
verb. Why not? When we try to visualize the structures in question,
the answer is immediate. Figure 5.4 illustrates what the different
combinations of subject placement and verb attraction look like in a
language where the heads consistently come at the end.

   Comparing the two diagrams shows that the subject comes first in
the sentence regardless of whether it attaches to the verb phrase or to

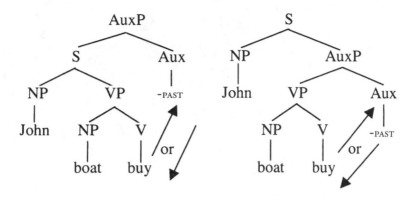

FIGURE 5.4   Subjects and Verbs in Head-Last Languages

the auxiliary phrase. Each diagram also shows that the inflected verb comes last regardless of whether the verb moves to the tense or the tense moves to the verb. The basic geometry of the clause is such that these parameters do not create word order variations in head-final languages. For all we know, there may be subtly different types of subject-object-verb languages, just as French and English are subtly different types of subject-verb-object languages; but the differences are so slight that they have not generally been noticed. Perhaps clever arguments can be constructed using adverbs or other elements to show that some head-final languages have verb hosts and others have tense hosts. Or perhaps there are no elements that could ever be in the right place to bring these differences to light. This is currently a subject of controversy among linguists. However it turns out, it is to the credit of the parametric analysis that it generates only those gross word orders that are actually attested. Within my chemical analogy, I suggested that we think of Welsh as an alloy of English, the result of adding a bit of carbon to the iron. The lack of discernible variations on Japanese is then like adding the same percentage of carbon to a different substance and not getting a useful alloy out of it because of the inherent properties of the atoms involved.

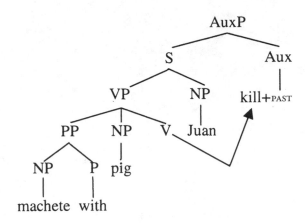

FIGURE 5.5   An Object-Subject-Verb Language (Warao?)

This style of reasoning can be pushed even to the remotest regions of Table 5.1. Consider verb-object-subject languages like Tzotzil and Malagasy, which constitute some 3 percent of the languages of the world. Do we expect minor variations on this word order because of the subject placement parameter and the verb attraction parameter? A moment with a bit of scratch paper should convince you that the answer is no. These languages are the exact mirror images of Japanese with respect to word order, so our secondary parameters have no obvious effect on them either. Subjects will come at the end, regardless of whether they are attached to the verb phrase or the auxiliary phrase, and verbs will come at the beginning, regardless of whether they attract tense or are attracted by it. Now consider object-verb-subject languages like Hixkaryana, which by Tomlin's count are 1 percent of the total (he found five among his 402). If these are truly the mirror images of English, do we expect minor variations on this word order because of the subject placement parameter and the verb attraction parameter? This time the answer is yes. A Hixkaryana-type language that also had low subject placement and tense auxiliaries as hosts should be Mirror Welsh—an object-subject-verb language (see Figure 5.5).

Of course, we should not find *many* such languages. The Hixkaryana word order is already very rare, and we would expect its Mirror Welsh variant to be rarer still—on the order of 25 percent of the total, just as Welsh-type languages are about 25 percent as common as English-type languages. Since Tomlin found five object-verb-subject languages, we might expect to find approximately one object-subject-verb language in his sample and on the order of ten in the 5,000-plus languages in the whole world. And perhaps we do; this could be where Warao fits in. Warao is a language spoken in Venezuela of which little is known, except that it is said to have object-subject-verb order:

Buare	isiko	ibure	hua	n-ae.
Machete	with	wild-pig	Juan	kill-PAST

'Juan killed a wild pig with a machete.'

Originally I said it was a good thing that the theory, which includes the verb-object constraint, predicted that no such language should exist. That was correct as a first-order approximation. Our second-order approximation says that such languages might not be impossible, but they would be the rarest of the rare, arising only when every parameter is set in just the right way. And that might be an even better prediction.

I confess that I have now ventured out to the fringe of knowledge—perhaps to the lunatic fringe. There are very few languages of the Hixkaryana and Warao types, and they are not among the world's best studied. Any conclusions about their structure are based on slender evidence and must be considered tentative. (Indeed, I suggest another analysis of Hixkaryana in the next chapter.) Nevertheless, the possibilities are worth discussing because they illustrate the conceptual issues that arise when one pursues the logic of parameters. In particular, they show how different parameter settings can interact in ways that sometimes derive new language types and sometimes do not. This point is important in the next chapter. We have also seen how a parametric theory might account not only for the types of word orders that are possible but also for the relative frequencies of those types.

————————————

The effects of the subject placement parameter and the verb attraction parameter on E-language are not as far-reaching as the other parameters we have seen. I have compared these parameters to loose rocks that do not cause landslides because they are near the bottom of a slope. But this does not mean they have *no* other effects; they are not necessarily at the very bottom of the slope. Sometimes a "minor" parameter like the verb attraction parameter makes its presence felt not only by positively influencing word order but by inhibiting other grammatical possibilities. Such cases show us, in a somewhat different way, how intricately interconnected natural languages are.

A good example of this inhibitory influence comes from the distribution of so-called serial verb constructions in languages of the world. Putting aside conjunctions using *and,* almost every English clause has exactly one subject and exactly one verb. But this one-to-one correspondence does not always hold in other languages. We saw in Chapter 2 that sentences in Spanish and Italian need not have a subject (other than the inflection on the verb; see Chapter 4). Serial verb constructions represent the opposite exception: They are sentences that contain more than one verb associated with a single subject and (at most) a single tense auxiliary. West African languages such as Edo and Yoruba are famous for having them. Here are two examples from Edo:

Òzó   ghá   lè   èvbàré   khiẹ̀n.
Ozo   will   cook   food   sell
'Ozo will cook the food and sell it.'

Òzó   ghá   suà   àkhé   dè.
Ozo   will   push   pot   fall
'Ozo will push the pot down [literally, so that it falls].'

English has no exact equivalent of these structures. The English translation of the first sentence requires a conjunction *and* and an object pronoun after the second verb, neither of which is visible in the Edo example. English speakers cannot say **Ozo will cook food sell*. For the second sentence, the English translation requires that the result be expressed as an adjective or preposition rather than as a second verb. The term *serial verb construction* indicates that verbs in these languages can come in series. Similar constructions are found in Southeast Asian languages such as Thai, Khmer, and Vietnamese. Since this possibility exists in some languages but not in others, I treat it as a parameter:

THE SERIAL VERB PARAMETER

Only one verb can be contained in each verb phrase (English, etc.).

or

More than one verb can be contained in a single verb phrase (Edo, Thai, etc.).

Consider now how the serial verb parameter might interact with the verb attraction parameter. Languages with a positive setting for the serial verb parameter will have sentences with only one tense auxiliary but two (or more) verbs. When there is no tense marker or when the tense marker is expressed as a separate word, this creates no particular difficulty. Imagine what happens, however, when the tense marker is expressed as an affix in a language like Welsh or French, where such tenses attract the verb to themselves. For single-verb sentences, this attraction is all well and good, but problems would arise in a serial verb sentence with more than one verb. This configuration can have no satisfactory outcome, just as it is difficult for a person with only one extra bed to have two friends stay overnight at the same time. There is no room inside the tense auxiliary for both verbs, since the tense affix can attach to only one verb at a time. If the tense auxiliary does not attract any verb, then it will be unpronounceable, left stranded with nothing to support it. Yet if the tense attracts only one of the two (or more) verbs to it, then the unattracted verb will have the linguistic equivalent of hurt feelings, making the sentence ungrammatical. As a result, serial verb constructions show up only in languages that either have no tense marking at all or that express tense as an independent word. Both the West African languages and the Southeast Asian languages have this property, but most European languages do not.

Edo is particularly interesting in this respect because it has exactly one tense that shows up as a suffix. This is a particular kind of past tense, expressed by the suffix *-re:*

Èvbàré   ọ̀ré   Òzó   lé-rè.
Food   FOCUS   Ozo   cook-PAST
'It's food that Ozo has cooked.'

(One special property of this tense is that it appears on a transitive verb only if the direct object has been moved to the front of the sentence as the focus. How and why this happens is not important here.) It is significant that this one tense is incompatible with serial verb constructions, which are otherwise common in Edo. Thus, there is no way to make a sentence like the following:

*Èvbàré   ọ̀ré   Òzó   lé-rè   khièn(-rèn).
Food   FOCUS   Ozo   cook-PAST   sell(-PAST)
'It's food that Ozo has cooked and sold.'

This sentence is bad, regardless of whether one puts the past tense suffix on the first verb, the second verb, or both. This is only a very minor inconvenience for Edo speakers, who have another past tense they can use in these situations. It is clear, however, that languages that depend more heavily on affixation to express tense (like French) are going to be languages that cannot have serial verb constructions. The two possibilities simply cannot coexist.

If all this is correct, then there should also be a negative interaction between the basic word order of a language and the possibility of having serial verb constructions. We have seen that languages with verb-subject-object order, such as Welsh, arise because the verb gets attracted to the tense auxiliary, thereby moving past the low-attached subject. Since this word order depends crucially on the verb attraction parameter (among other things), and since verb attraction is incompatible with verb serialization, we predict that serial verb constructions will never be possible in verb-subject-object languages—even though they are possible in otherwise similar subject-verb-object languages.

Data reported by Eric Schiller show that this prediction is correct. His most striking evidence comes from the Khmer languages. Historically, the Proto-Khmer language from which all the modern languages have descended had the Welsh-style verb-subject-object word order. This order has been maintained in some of the modern languages, such as Ravua. Others, like Modern Khmer (the primary language of Cambodia), switched to English-like subject-verb-object order somewhere along the way. This development can be thought of as a change in the setting of the verb attraction parameter over time. What is striking is that the Khmer languages that have developed subject-verb-object word order have also developed serial verb constructions. Those that have maintained the original verb-subject-object order have resisted this trend, in spite of the peer pressure that comes from many of the surrounding languages having this feature. This supports the idea that verb attraction and verb serialization are incompatible features in a language's recipe. Including both would be like adding baking soda directly to vinegar while cooking: The two react to produce a fizzy mess of little practical use.

Thus, parameters can interact with each other so as to set limits on the overall properties of a language. Such interactions seem quite common. They ensure that although all languages have complexities, no language can be complex in every way at once.

The variations we have seen in this chapter so far can all be thought of as linguistic alloys, in which a "minor" parameter adds its influence to a "major" one. Welsh is a variant of an English-style language created by adding a dash of verb attraction parameter; Edo is a variant of an English-style language created by adding a pinch of serial verb parameter. Yet some languages cannot comfortably be classified as variants of one main type. Instead, they seem to be true intermediates, combining characteristics of two major types into a single language. Within the chemical analogy, such languages can be thought of as compounds, formed by a deeper fusion of two linguistic atoms.

An excellent example of a compound language is Chichewa, a language spoken in Malawi. (Chichewa is a Bantu language, related to Swahili, Africa's best-known language; Swahili also has the properties I discuss.) Chichewa can be described as a blend of an English-style language and a Mohawk-style language. In Chapter 4 we saw that verbs in Mohawk always agree with both their subjects and their objects, whereas verbs in English for the most part do not agree with either. (There is some residual agreement with subjects in English, carried over from an earlier stage of the language, but it has little impact on syntax.) Chichewa verbs always agree with their subjects, and they can agree with their objects as well. Object agreement is not required in Chichewa, however, the way it is in Mohawk:

Njuchi	zi-na-luma	alenje	[without object agreement].
Bees	they-bit	hunters	*or*
Njuchi	zi-na-*wa*-lum-a	alenje	[with object agreement *wa*].
Bees	they-bit-them	hunters	

'The bees bit [i.e., stung] the hunters.'

In English any clause that has a transitive verb must have both a subject noun phrase and an object noun phrase, whereas Mohawk requires neither. In Chichewa the subject noun phrase is never necessary. The object noun phrase can be left out, but only if there is an object agreement:

| Zi-na-(wa)-luma | alenje | [subject omitted]. |
| They-bit-them | hunters | |

'They bit the hunters.'   *or*

		[object omitted].
Njuchi	zi-na-*wa*-lum-a	
Bees	they-bit-them	

'The bees bit them.'   *but NOT*

*Njuchi   zi-na-lum-a   [object and object agreement both omitted].
Bees   they-bit
'The bees bit.'

Finally, in English the order of the subject, verb, and object is rather rigidly fixed, whereas in Mohawk it is very free, depending on what the speaker wants to emphasize. In Chichewa sentences with no object agreement, the position of the subject is free: It can appear at the beginning of the sentence or the end. But the position of the object is fixed: It has to be immediately after the verb, just as in English. One thus finds the following two word orders but no others:

Njuchi   zi-na-lum-a   alenje.   [subject-verb-object]
Bees   they-bit   hunters
'The bees bit the hunters.'   *or*
Zi-na-lum-a alenje   njuchi.   [verb-object-subject]
They-bit   hunters   bees

Yet when the verb contains an object agreement, Chichewa word order is just as free as Mohawk. All of the six logically possible arrangements of subject, object, and verb are grammatical, with the same meaning of 'The bees stung the hunters':

Njuchi   zi-na-wa-lum-a   alenje.   [SVO]
Bees   they-bit-them   hunters   *or*
Zi na wa-lum-a   alenje   njuchi.   [VOS]
They-bit-them   hunters   bees   *or*
Zi-na-wa-lum-a   njuchi   alenje.   [VSO]
They-bit-them   bees   hunters   *or*
Alenje   njuchi   zi-na-wa-lum-a.   [OSV]
Hunters   bees   they-bit-them   *or*
Alenje   zi-na-wa-lum-a njuchi.   [OVS]
Hunters   they-bit-them bees   *or*
Njuchi   alenje   zi-na-wa-lum-a.   [SOV]
Bees   hunters   they-bit-them

Chichewa thus combines Mohawk-like and English-like properties. It seems to have a Mohawk-like mode with object agreement and a second mode (without object agreement) in which it treats its direct object essentially as English does. This is what I mean by a linguistic compound.

Compound languages of this type seem to be reasonably common. In addition to all of the Bantu languages, other languages with obligatory subject agreement, optional object agreement, and head-first tendencies include the Nilo-Saharan language Lango, the Indonesian language Selayarese, and perhaps the Chilean language Mapuche. Even Romance languages such as Spanish and Italian could be thought of in this way, if one takes the "weak" object pronouns that attach to the verb in these languages to be the same thing as object agreement in Mohawk and Chichewa.

Other kinds of compound languages exist as well. For example, the Slave (pronounced "slay-vee") language spoken in the Yukon Territory is a blend of Japanese and Mohawk in much the same way that Chichewa is a blend of English and Mohawk. In some Slave sentences, the verb agrees only with the subject, not the object. In that case the object must be present, and it comes between the subject and the verb, as in Japanese:

Li	'ehkee	kayihshu.
Dog	boy	it-bit

'The dog bit the boy.'

In other sentences the verb bears an object agreement *(ye)* as well. When this happens, the object's position is less constrained. It can come at the beginning of the clause, before the subject:

'ehkee	li	kay*e*yihshu.
Boy	dog	it-bit-him

'The dog bit the boy.'

Or if the prefix *ye* is present, the object can be left out entirely:

Li   kayeyihshu.
Dog   it-bit-him
'The dog bit him.'

These are options that are characteristic of Mohawk-type languages. These last two sentences, however, are ungrammatical if the object agreement *ye* is removed from the verb. This blend of properties also seems to be rather common. Other languages that have it include the South American language Imbabura Quechua, some New Guinean languages, and perhaps Abkhaz, spoken in the Caucasian Mountains.

It is important to realize that not every compound language one can readily imagine constitutes a possible human language. Chichewa and Slave are alike in that they both treat their subjects in the Mohawk way and their objects in the English/Japanese way. One can just as well imagine the opposite blend. This would be a language in which the verb always agreed with the object, so that the object could be omitted or put on either edge of the sentence. But the verb would not have to agree with the subject, which would have to be present and in its usual position before the main verb or tense auxiliary. Such a language, which we could call Reverse Chichewa, would have the following range of sentences:

Bees	bit-him	John	[ordinary sentence]	*or*
Bees	bit-him		[object omitted]	*or*
John	bees	bit-him	[object relocated]	*but not*
*Bit-him	John		[subject omitted]	*nor*
*Bit-him John bees,		*Bit-him bees John, etc. [subject relocated]		

Reverse Chichewa might also allow subject agreement as an option, and when it was used the subject noun phrase could be left out and word order would be freer. No such language is known to exist. Nor is there a Reverse Tzotzil, a hypothetical language that would be just like Reverse Chichewa but with the unagreed-with subject fixed at the end of the clause rather than the beginning.

These restrictions on possible blends make linguistic compounds interestingly like their chemical namesakes. In linguistics, as in chemistry, some combinations of atoms form stable compounds, but others do not. Just as the atomic theory of chemistry has a subtheory of valence that explains what compounds are possible, so the parametric theory of language should explain the possible blends.

There is no consensus among linguists about how best to handle cases like these. Within the parametric theory of language, one might proceed in at least three plausible ways, depending in part on which view of parameters one subscribes to. Below, I mention all three briefly and then say which one I am inclined to prefer and why, fleshing it out in a bit more detail.

The first and perhaps the most obvious possibility is to carve up the head directionality parameter and the polysynthesis parameter into smaller units. So far I have stated these large-scale parameters in the most general possible way: The head directionality parameter applies to the construction of *every* phrase in the language, and the polysynthesis parameter says that *every* participant in the event must be expressed on the main verb. Because of this universal cast, these parameters apply in a uniform way to both subjects and objects. For languages such as English, Japanese, and Mohawk, this is a very good thing. But now we find that we also need a way to describe languages like Chichewa and Slave, which treat subjects and objects differently. We might accomplish this by saying that the polysynthesis parameter is not a single recipe statement but rather a family of conceptually similar but narrower recipe statements. One of these statements would say that the subject must be expressed on the verb, another would say that the object must be expressed on the verb, and so on:

THE SUBJECT POLYSYNTHESIS PARAMETER

The subject of a verb must be expressed in that verb (Mohawk, Chichewa, Slave).

or

The subject of a verb need not be expressed inside that verb (English, Japanese).

THE OBJECT POLYSYNTHESIS PARAMETER

The object of a verb must be expressed inside that verb (Mohawk).

or

The object of a verb need not be expressed inside that verb (English, Japanese, Chichewa, Slave, etc.).

Consistent languages like English and Mohawk can still be expressed in such a system; the only inelegance is that it takes two decisions to get these languages rather than one. English would be a language in which neither the subject nor the object needs to be expressed in the verb; Mohawk would be a language in which both have to be expressed in this way. The payoff is that there is now a way to describe Chichewa and Slave: These are languages that are required to have an expression of the subject on the verb but not an expression of the object. The head directionality parameter could also be split up into a family of smaller-scale parameters, each applying only to certain kinds of phrases.

The second strategy for explaining compound languages like Chichewa and Slave focuses not on avoiding conflicts between parameters but on resolving them. This strategy would maintain both the polysynthesis parameter and the head directionality parameter in their full generality but would add a principled way of negotiating the competing demands they put on a single structure. This is the natural approach within optimality theory, described briefly at the end of Chapter 3. Optimality theory claims that all principles hold in all languages, and the differences among languages come not from which principles are selected but rather from which are given priority. This view has some initial promise. We can think of the head directionality parameter as conflicting with the polysynthesis parameter over the matter of how the direct object of a verb should be expressed. The positive setting of the polysynthesis parameter wants the object to be agreed with, and this has the automatic effect of dislocating an overt noun phrase related to the object to the edge of the clause (or so I have been assuming). But the "head first" setting of the head directionality parameter wants the object to come after the verb, forming a verb phrase with it. The "head last" setting of the head directionality parameter, conversely, wants the object to come before the verb. Mohawk, Chichewa, and Slave can be seen as the results of different choices in which of these principles gets its own way. If the polysynthesis parameter is given the highest priority, the result is Mohawk; if the head-first parameter wins out, the result will be Chichewa; if the head-last parameter wins, the result will be Slave. Notice, however, that the differences among these three languages all surround the treatment of the direct object; they all handle subjects in the same way, as being agreed with and dislocated. This point of similarity follows elegantly from the fact that, as we saw in Chapter 3, *the head directionality parameter says nothing about where the subject should go.* (The placement of the subject is determined by two additional parameters: the subject side parameter and the subject placement parameter.) Thus, there is no inherent conflict between the polysynthesis parameter and the head directionality parameter over the positioning of the subject. Only the polysynthesis parameter has views about this, so its views carry the day. This is how the optimality theory approach can potentially explain why languages like Chichewa are possible but languages like Reverse Chichewa are not.

The third strategy for dealing with compound languages is to allow that parameters may have more than two settings. Suppose there is a third possible setting for the polysynthesis parameter:

The Extended Polysynthesis Parameter

All participants of an event must be expressed on the verb (Mohawk).

or

Any participant of an event may be expressed on the verb (Chichewa, Slave).

or

No participant of an event is expressed on the verb (English, Japanese).

The second of these three options is the new one, intended to make room for Chichewa and Slave; we can call it the "optional" setting of the polysynthesis parameter, or the optional polysynthesis parameter for short. This strategy is a close conceptual cousin to the first one since both make new recipe options available (albeit in different ways). It has the advantage of automatically creating the right number of languages. In principle, we predict six languages to result from the two settings of the head directionality parameter being crossed with three settings of the polysynthesis parameter. In practice, however, the setting of the head directionality parameter has little or no effect on a Mohawk-type language because subjects and objects never form phrases with the verb in Mohawk-type languages. The third strategy thus predicts precisely the five different kinds of languages we have been discussing, no more and no less, as shown in Table 5.2.

Although all three strategies succeed in explaining why objects are treated differently from subjects in some languages, they also share a common problem. Basically, they are too permissive. The same rules that allow a language to have agreement with the subject but not the object (as found in Chichewa and Slave) also allow a language to have agreement with the object but not the subject. This is an undesirable result given that no Reverse Chichewa languages are known. This problem arises most clearly in the first strategy, which fragments the parameters. On this view, it should be possible for a language to pick the negative setting of the subject polysynthesis parameter and the positive setting of the object polysynthesis parameter. These settings would produce Reverse Chichewa immediately.

The second strategy, based on the optimality theory idea of setting priorities among conflicting constraints, seemed at first glance

TABLE 5.2   Simple and Compound Languages

	Head-First Languages	Head-Last Languages
No expression on verb	English, Edo, etc.	Japanese, Lezgian, etc.
Optional expression on verb	Chichewa, Seyalarese, etc.	Slave, Quechua, etc.
Required expression on verb	Mohawk, Mayali, etc.	

to avoid this consequence. But that was only because our comparison set was artificially small. We considered only Mohawk, Slave, and Chichewa, all of which have subject agreement and dislocation of subject noun phrases. Not all languages have these properties, however: In particular, English and Japanese do not. Consequently, a fuller description must include some other (as yet unknown) constraint that takes priority over the polysynthesis parameter in English and Japanese, preventing subject agreement. Once this constraint enters the picture, strategy two also permits Reverse Chichewa. Reverse Chichewa would arise when the mystery constraint is ranked above the polysynthesis parameter (preventing agreement with subjects), and the polysynthesis parameter is ranked above the head directionality parameter (requiring agreement and dislocation of objects).

Therefore, neither of these strategies presently constitutes an adequate theory of linguistic compounds. We would not be impressed by chemists who told us that their theory explained how water can be created out of hydrogen and oxygen, if their theory also predicted that hydrogen and oxygen could form table salt. Neither should we be impressed with a linguistic theory that predicts both Chichewa and Reverse Chichewa. A satisfying theory must be able not only to describe everything that happens but also to explain why some things never happen.

The third strategy has a slightly different problem. It does not predict the existence of extra languages, as shown in Table 5.2. If anything, its problem is even more embarrassing: It does not get the properties of Chichewa and Slave exactly right. As it stands, the optional polysynthesis parameter says that agreement with the subject is not any more required than agreement with the object in these languages. But this seems to be false: Subject agreement *is* required in these languages. Strategy three gives us the right number of languages, but those languages have the wrong qualities.

As embarrassing as this fault is, there is some reason to think that the problem with strategy three might be the easiest to patch up. We can gain some leverage on the issue by looking at the factors that condition the use of object agreement in languages in which it is optional. The effect of optional object agreement is rather consistent. In language after language, object agreement is likely to be used when the direct object refers to a person or a definite object known to both the speaker and the hearer. In contrast, object agreement is not used when the direct object refers to an indefinite inanimate. This is nicely seen in Swahili, which has the same blend of Mohawk and English properties as Chichewa:

| Juma | a-na-*wa*-penda | watoto. | [agreement *(wa)* with animate object] |
| Juma | he-likes-them | children | |

'Juma likes children'; 'Juma likes the children.'

| Juma | a-li-*li*-kamata | gitara. | [agreement *(li)* with definite object] |
| Juma | he-grabbed-it | guitar | |

'Juma grabbed the guitar.'

| Juma | a-li-kamata | gitara. | [no agreement with indefinite, inanimate object] |
| Juma | he-grabbed | guitar | |

'Juma grabbed a guitar.'

Roughly the same generalization holds for other languages, although the details vary somewhat. Thus, the master recipe for language should include a principle that guides speakers in the use of optional agreement markers—something along the following lines:

THE AGREEMENT PRINCIPLE

If agreement with a noun phrase X is not required, use the agreement to show that noun phrase X is animate and/or definite in its reference.

Now, it is a significant characteristic of subjects that they are much more likely to be people or specific things than objects are. This tendency can be seen in English: The subject of a transitive sentence is usually a person (or an animal) or a noun phrase with the article *the* (or its equivalent), as any reader with a newspaper handy can confirm. In part this is because subjects typically express the agents in events, and agents are usually animate. In English this is only a tendency, since sentences with an indefinite and inanimate subject are also possible. For example, the sentence *A rock must have hit him on the head* is possible, even though the rock in question is not animate and the speaker does not have a particular identifiable one in mind. Other languages avoid this kind of sentence more vigorously; they might use a passive like *He was hit on the head by a rock* instead. Navajo and Southern Tiwa do this. Now, if it is true that subjects are usually or always definite or animate, then the agreement principle will always tell speakers who have a choice to use agreement with the subject, even though agreement is optional in theory. This, then, is a possible way of plugging the hole in the third strategy.

One additional consideration points in favor of the third solution. Both the parameter-splitting approach and the parameter-ranking approach predict that blended languages like Chichewa and Slave should treat their subjects *exactly* as Mohawk does. But there is one respect in which this is not true. In Chapter 4 I pointed out that the verb in both Spanish and Mohawk always agrees with its subject *if* the verb bears a tense marking. Spanish, however, also has infinitive forms of the verb that it uses in various contexts as part of a subordinate clause. These infinitival verbs do not show agreement with a subject. Mohawk and the other true polysynthetic languages have no such infinitival form; every verb must be marked for subject agreement, regardless of its tense or syntactic position. I took this to be a reflection of the polysynthesis parameter, a sign that Mohawk is more serious about needing subject agreement than Spanish is. It turns out that the blended languages are like Spanish rather than Mohawk in this respect. Chichewa, for example, has infinitival verbs, marked by the prefix *ku-*. Such verbs do not agree with their understood subjects, the way other verbs do in Chichewa:

Asilikálí	s-a-ngath-é	*ku-gwira*	chigawéngá	ichi.
Soldiers	they-cannot	to-catch	terrorist	this

'The soldiers cannot catch this terrorist.'

The same is true for other "optional agreement" languages such as Lango and Quechua (Slave is an exception). The first two strategies have no obvious explanation for why Chichewa-type languages tend to differ in this respect from Mohawk-type languages. But this difference is not at all surprising from the perspective of the optional polysynthesis parameter. On the contrary, this parameter explicitly denies that subject agreement is required in languages like Chichewa. When we add infinitives to the picture, then, we see that subject agreement is not necessary in Chichewa-type languages after all. The factors that determine whether an object agreement will be used may still turn out to be at least partly different from the factors that determine whether a subject agreement will be used. I see no reason, however, to doubt that this last phase of the problem can be filled in appropriately, completing the parametric account of this type of linguistic compound.

Life is dull if you have only one kind of stuff. It is usually not much fun to paint a picture with one pigment, but having three that you can mix makes it appealing. There are few baked goods that you can make with only flour, and the chemistry of one-element molecules is of limited interest. The same is true in linguistics: One can explain only so much with just one parameter. This chapter has looked at some of the nuances that come into a parametric theory when one begins to consider the various ways that parameter settings can interact. Sometimes a "minor" parameter creates two subtypes of language, such as the Welsh type and the English type. Sometimes one parameter prevents another from being expressed, as verb movement blocks serial verb constructions in Khmer languages. Sometimes universal principles skew the way an optional feature of language is deployed, creating an otherwise unexpected asymmetry, as the agreement principle affects the optional polysynthesis parameter in Chichewa-type languages. Finally, sometimes a parameter is simply irrelevant to a particular type of language, as the verb attraction parameter is to Japanese. These interactions give a fuller picture of the true chemistry of human languages. The properties of these interactions can also provide a guide to fulfilling an ultimate goal: the construction of a linguistic equivalent to chemistry's periodic table of the elements.

6

Toward a Periodic Table of Languages

I N CHAPTER 2 I COMPARED OUR CURRENT UNDERSTANDING of linguistic diversity to the stage chemistry was in during the middle of the nineteenth century, just before Mendeleyev presented his periodic table of the elements. The basic ingredients of the theory have been identified: parameters in linguistics, atoms in chemistry. The basic logic of how such ingredients can be used to address the discipline's fundamental questions is in place: how parameters explicate the Code Talker paradox in linguistics, how atoms account for the mutability of substances in chemistry. Furthermore, a reasonable number of the actual elements are known. We have specific formulations of the head directionality parameter, the polysynthesis parameter, the verb attraction parameter, and a few others, much as nineteenth-century chemists were acquainted with carbon, nitrogen, oxygen, and some other elements. In all, about 60 percent of the elements were known when Mendeleyev constructed the periodic table, including most (but not all) of the common ones. Our knowledge of large-scale parameters is probably at about the same point, judging from the percentage of statements in descriptive grammars that can be treated in terms of extant linguistic theory. Finally, the mechanisms by which linguistic parameters interact to create a limited diversity of languages have been worked out, just as nineteenth-century chemistry had worked out the valence properties of atoms.

For chemistry, the next step was to assemble these pieces into a comprehensive vision of the chemical world. The time may be growing ripe for linguistics to take a similar step—to construct a "periodic table of languages." That step has not yet been taken; no such grand scheme has dawned on the linguistics world. Nevertheless, from the glimmers we do see we may make some guesses at what such a table might be like. The "table" I present aspires to be like Beguyer de Chancourtois's "telluric screw" in the history of chemistry (see Figure 2.1) or like John the Baptist in the history of religion: It will not be a full and mature expression of the truth but a foreshadowing, intended to get people ready to recognize and appreciate the full truth when it comes.

———————

The periodic table as it is now usually presented is shown in Figure 6.1. Anything worth comparing to it should have two key features. First, it should be *complete*. Part of the glory of the periodic table of the elements is that every element in the universe is there somewhere. No matter what exotic material one comes across—antimony or strontium or ytterbium—it has its own box in the table, complete with chemical symbol, atomic number, and other information. The parametric theory of linguistics is built on the hypothesis that all grammatical differences among languages result from the interplay of a finite number of discrete factors. If this is correct, then those parameters should also be expressible in an exhaustive list. A periodic table of languages would be such a list, so that whatever exotic grammatical feature one might come across—a serial verb construction or an incorporated noun or an ergative case marker—it would be somewhere on the table of languages.

The second distinctive feature of the periodic table is that it is *systematic*. Not only are all the elements included, but they are placed in a natural order. The atomic number of each element is exactly one more than the element to its left, and the chemical valence of each element is the same as the one just above it. Thus, the arrangement of the

	IA	IIA	IIIB	IVB	VB	VIB	VIIB	VIII	VIII	VIII	IB	IIB	IIIA	IVA	VA	VIA	VIIA	0
1	1 H																	2 He
2	3 Li	4 Be											5 B	6 C	7 N	8 O	9 F	10 Ne
3	11 Na	12 Mg											13 Al	14 Si	15 P	16 S	17 Cl	18 Ar
4	19 K	20 Ca	21 Sc	22 Ti	23 V	24 Cr	25 Mn	26 Fe	27 Co	28 Ni	29 Cu	30 Zn	31 Ga	32 Ge	33 As	34 Se	35 Br	36 Kr
5	37 Rb	38 Sr	39 Y	40 Zr	41 Nb	42 Mo	43 Tc	44 Ru	45 Rh	46 Pd	47 Ag	48 Cd	49 In	50 Sn	51 Sb	52 Te	53 I	54 Xe
6	55 Cs	56 Ba	57 La	72 Hf	73 Ta	74 W	75 Re	76 Os	77 Ir	78 Pt	79 Au	80 Hg	81 Tl	82 Pb	83 Bi	84 Po	85 At	86 Rn
7	87 Fr	88 Ra	89 Ac	104 Rf	105 Ha	106 Unh	107 Uns											

6	58 Ce	59 Pr	60 Nd	61 Pm	62 Sm	63 Eu	64 Gd	65 Tb	66 Dy	67 Ho	68 Er	69 Tm	70 Yb	71 Lu
7	90 Th	91 Pa	92 U	93 Np	94 Pu	95 Am	96 Cm	97 Bk	98 Cf	99 Es	100 Fm	101 Md	102 No	103 Lr

FIGURE 6.1   The Periodic Table—a Standard Presentation

elements in the table communicates essential properties to the informed observer at a glance. Ideally, the same should be true of a periodic table of languages. Not only should each parameter be listed, but the parameters should be presented systematically, in a way that expresses truths about their inherent nature and the relationships among them.

Of these two goals, systematicity and completeness, it makes sense for us to seek systematicity first, for several reasons. First, completeness can be dull. There are few things less exciting than a long, unanalyzed list. (Who really cares about ytterbium anyway?) Second, the discovery of a natural order is often facilitated by focusing on a few things at a time. Fewer items have many fewer combinatorial possibilities, and sometimes a wealth of detail distracts from the deeper rhythms. Mendeleyev himself included only thirty-two elements, roughly half of the number then known, in his first table. Third, discovering the natural order of things can contribute to the quest for completeness; missing elements show up as gaps in the system only after the system as a whole has been discerned. One of Mendeleyev's great successes was predicting the existence of germanium, scandium, and gallium because he found that his periodic table worked best if he left blank boxes near silicon, boron, and aluminum. Perhaps a similar achievement is possible in linguistics.

For these reasons, I begin the task of imagining a periodic table of languages by looking for a suitable way to display the handful of parameters that we have already discussed. Then I briefly introduce a few more parameters into the mix, to see how well they fit into the proposed system and to get a little closer to the ideal of completeness.

––––––––––––––––

What would be a useful order for the parameters that define language differences? So far we have seen eight specific parameters: the null subject parameter (Chapter 2), the head directionality parameter (Chapter 3), the subject side parameter (Chapter 3), the polysynthesis parameter (Chapter 4), the subject placement parameter, the verb attraction parameter, the serial verb parameter, and the optional polysynthesis parameter (all in Chapter 5). These, then, are our dramatis personae. But since linguistics is not a movie, order of appearance is not the best way to present the cast.

Nor can we expect to imitate Mendeleyev's table too closely. For all of the conceptual parallels between the role of atoms in chemical theory and the role of parameters in linguistic theory, atoms and parameters are fundamentally different things. Atoms are little chunks of matter, whereas parameters are parts of some kind of mental knowledge structure that constitutes our linguistic abilities. They are like steps in a recipe or blocks of code in our internal programming for language. At the heart of Mendeleyev's table was a correlation between two physical properties of the elements: their atomic weights and their chemical valence. Just why these two properties should be correlated did not become clear until the early twentieth century, when physicists began to uncover the internal structure of atoms. But the existence of this correlation is the reason why the chemical elements can be displayed so nicely in a two-dimensional table, with rows and columns— why Mendeleyev's fascination with solitaire games happened to lead him to a format that proved so apt. These physical considerations will not be directly relevant to the parameters, and in this respect our overarching analogy has reached its limits. There is not even any guarantee that the natural ordering of parameters will be two-dimensional. In the strict sense, we may not be looking for a table at all. Rather, given the sorts of things parameters are, their natural ordering is more likely to resemble a flowchart for an algorithm or procedure.

In the previous chapter we already saw the signs of an important kind of logical ordering. Some parameters have a much greater impact on the form of an E-language than others do. Indeed, some parameters end up having no perceptible effect at all on the E-language, when some other parameter has been set in a way that makes the first one irrelevant in practice. In Chapter 5 I compared parameters to loose rocks on a mountain slope: All rocks have the same basic structure, but those located near the top of the slope may cause an avalanche of effects, whereas those near the bottom have no such opportunity. This analogy suggests the possibility of a useful order for the parameters: They could be presented in terms of their placement on this metaphorical slope of cause and effect. This placement need not be seen in temporal terms, as reflecting the order in which parameters are used as our minds process language. (There is probably no stable temporal ranking to the parameters because we undoubtedly use them in different orders as we perform different linguistic tasks.) Rather, it would be a purely logical order, in which parameters are ranked by their power to affect one another and their potentials for rendering each other irrelevant.

To see a specific example of how this works, consider the logical relationship between the polysynthesis parameter and the head directionality parameter. In terms of their actual statements, these parameters seem to be concerned with different matters. The polysynthesis parameter determines whether the participants of an action have to be represented on the verb that expresses the action, whereas the head directionality parameter determines the order in which words are assembled into phrases. But there is an important relationship between them in practice. One of the consequences of the polysynthesis parameter in languages like Mohawk, where it is set positively, is that agreement markers are included on the verb as expressions of the verb's subject and object. This causes full noun phrases that are associated with these roles to be dislocated to the edge of the clause. Such noun phrases do not combine with the verb to form verb phrases, as they do in other languages. The head directionality parameter is in practice irrelevant to the verb-object relationship in Mohawk (see Table 5.2). In this sense, the polysynthesis parameter "bleeds" the head directionality parameter. Moreover, since the polysynthesis parameter applies to *all* the participants in an event, not just the direct object, similar effects are found in other kinds of phrases. If this reasoning is carried through consistently, the head directionality parameter could simply be irrelevant to Mohawk-style languages because the kinds of grammatical configurations it regulates never arise. As powerful as it is in creating the extensive differences between English-style and Japanese-style word order, then, the head directionality parameter is not the most powerful grammatical force. It can be rendered impotent by one setting of the polysynthesis parameter. The polysynthesis parameter thus has a kind of logical priority, determining whether the head directionality parameter can express itself or not. This is a principled reason for ranking the polysynthesis parameter above the head directionality parameter. This ranking could be expressed visually in a table of languages by the convention of placing the polysynthesis parameter higher on the page than the head directionality parameter.

Based on the other examples we have seen, it is not unusual for one parameter to have logical priority over another in this way. For example, we saw in Chapter 5 that the verb attraction parameter has important effects in head-first languages such as English. On the one hand, it creates the Welsh and French variants of English-style word order, in which verbs come before adverbs and perhaps before subjects. On the other hand, the same parameter has no appreciable effect on head-final languages like Japanese. It so happens that in head-final languages the tense marker and the verb are already next to each other, so it doesn't matter much which one moves. This means that the head directionality parameter has logical priority over the verb attraction parameter.

Similarly, we saw in Chapter 5 that the serial verb parameter, which says that more than one verb can be contained within a verb phrase, gets a chance to express itself only within languages in which the tense auxiliary does not attract the verb. Thus, the verb attraction parameter has logical priority over the serial verb parameter.

This suggests that the relationship of logical priority is pervasive enough and important enough to be worth building a table of languages around. In such a table, the head directionality parameter would appear below the polysynthesis parameter but above the verb attraction parameter, which would in turn appear above the serial verb parameter. The ranking principle can be stated in a general form as follows:

Parameter X ranks higher than parameter Y if and only if Y produces a difference in one type of language defined by X, but not in the other.

Not all parameters will be ordered with respect to each other by these considerations. Sometimes one parameter will have an equivalent effect on both of the language types defined by another parameter, and vice versa. When this happens, the combination of two parameters that have two settings each gives a total of four distinct language types. The two parameters are then logically independent— probably because they characterize noninteracting aspects of language. In my analogy these parameters would sit side by side on the mountain slope. An actual case in point is the head directionality parameter and the optional polysynthesis parameter, which says that all participants in an event can optionally be expressed in the verb that denotes the event. The original polysynthesis parameter bleeds the head directionality parameter because it completely prevents words from forming phrases of the relevant kind. The optional polysynthesis parameter is not so forceful: It creates possibilities but not requirements. A noun phrase may be dislocated in such a language, but it may also appear as a phrase together with the verb. Whenever this second option is taken, the head directionality parameter determines whether the verb comes first or last in the phrase. All four of the logically possible ways of setting these two parameters thus generate clearly distinct E-languages, as shown in Table 6.1. These two parameters are therefore not ordered with respect to each other.

In other cases we may need more information before we can confidently rank a pair of parameters. An example is the interaction between the head directionality parameter and the possibility (entertained in Chapter 3) of a subject side parameter. Recall that the placement of subjects inside clauses is not determined by the head directionality parameter because subjects are phrases that combine with predicates, which are themselves phrases. Thus, we considered the possibility of adding an additional parameter to determine whether the subject came at the beginning of the clause, as in English, Edo, and Vietnamese, or at the end, as in Tzotzil and Malagasy. Suppose such a parameter exists. The next question is whether this subject side parameter is ranked with respect to the head direction-

TABLE 6.1   Head Directionality and Optional Polysynthesis

	Head First	*Head Last*
No expression on verb	English, Edo, Vietnamese	Japanese, Lezgian
Optional expression on verb	Chichewa, Seyalarese, Lango	Slave, Quechua, Abkhaz

ality parameter. The evidence on this point is equivocal and needs further interpretation. By hypothesis, the subject side parameter is relevant to head-initial languages, distinguishing English from Tzotzil. The crucial question is whether the subject side parameter also makes a difference in head-final languages. If it does, then the two parameters are unordered; if it does not, then the head directionality parameter has logical priority (see Table 6.2).

Head-final languages with sentence-final subjects have occasionally been reported: Hixkaryana, a Carib language spoken in Amazonia, is the best-described one. This seems to support the independence of the two parameters. In Chapters 3 and 5, I tentatively assumed that these reports could be taken at face value and used this to illustrate the logic of the parametric analysis. But there are reasons to reevaluate this decision. It may have been *too* easy to give an analysis for object-verb-subject languages. If they are really the result of setting two simple word order parameters, then one would expect that something like 25 percent of the world's non-polysynthetic languages should have object-verb-subject order. That is far from true: This particular order has been adduced in at most five less-studied languages, all of the Carib family, so 0 percent seems like a more accurate estimate of the proportion of these languages than 25 percent. But this is exactly the kind of no-win situation in which linguists practice their name-calling: Typologists accuse theorists of suppressing the data, and theorists accuse typologists of being easily deceived by any exotic-sounding rumor.

TABLE 6.2   Word Order and Subject Location

	Head Initial	Head Final
Subject first	Subject-verb-object: English, Edo, Vietnamese	Subject-object-verb: Japanese, Lezgian
Subject last	Verb-object-subject: Malagasy, Tzotzil	Object-verb-subject? (Hixkaryana?)

There are also details in the description of Hixkaryana that add to the suspicion. Indirect objects, for example, consistently come after the subject:

Otweto	yɨmyakonɨ	rohetxe	totokomo	wya.
Hammock	gave	my-wife	people	to

'My wife used to give hammocks to the people.'

This is not the order we would predict on the simple analysis assumed in the previous chapters; rather, the indirect object should, like the direct object, come before the verb, as in Japanese. There is no easy way to get this word order that is consistent with the verb-object constraint. Richard Kayne takes this as evidence that object-verb-subject languages are not created by special word order parameters at all but instead by a kind of movement akin to the question movement discussed in Chapter 2. Hixkaryana sentences could start out as ordinary subject-object-verb structures, with the direct object and the verb forming the smallest phrase (as usual). This verb phrase then moves to the front of the sentence, creating Hixkaryana's distinctive word order:

my-wife people-to [hammock gave]   (initial order)

→

[hammock gave] my-wife people-to —(derived order)

(The brackets here enclose the smallest verb phrase, which is the moved item.) Confirmation for this view comes from the occasional appearance of subject-object-verb order in nonfinite embedded clauses:

Ro-wy   *wewe*   *yamatxhe,*   ɨtehe   harha   owo   hona.
Me-by   tree   after-felling   I-go   back   village   to
'After I fell the tree, I will go back to the village.'

This is what we expect if verb phrase fronting cannot take place in this environment. More generally, if verb phrase movement is relatively rare (as seems to be the case), Kayne's proposal explains why languages like Hixkaryana are rare. Thus, the few object-verb-subject languages that exist do not necessarily show that the subject side parameter is independent of the head directionality parameter. In what follows, I assume that Kayne's analysis of Hixkaryana is correct and that the head directionality parameter outranks the subject side parameter. But this is not crucial to the account.

———————

The logical relations among parameters thus do not give a complete ordering but at best a partial one, in which priority is defined for some pairs but not others. A total ordering is a one-dimensional object; it is a line in which every point comes either before or after every other point. A partial ordering is two-dimensional, with points not only above and below each other but also side by side. Our tentative "table of languages," then, may be represented as a chart in two dimensions. This is a satisfying result, if only because two-dimensional structures are convenient for printing on paper.

Before presenting my first draft of such a table, I must offer tentative assessments of where the other parameters we know about fit in. I have not yet mentioned two parameters from the earlier chapters: the subject placement parameter (from Chapter 5) and the null subject parameter (from Chapter 2). The subject placement parameter seems to be fairly low on the list, roughly on a par with the verb attraction parameter. It concerns whether subject noun phrases are combined with the verb phrase or with the auxiliary phrase, and it works together with the verb attraction parameter in head-initial languages to create verb-subject-object order in languages like Welsh and Zapotec. It is clearly subordinate to the head directionality parameter and the subject side parameter because its effects are seen only in the subject-verb-object family of languages. Whether or not it is also subordinate to the verb attraction parameter depends on an empirical question. In Chapter 5 I discussed three types of languages that are created by these two parameters: the Welsh type, the French type, and the English type. One might logically expect a fourth type of language to exist—one in which the subject comes between a tense auxiliary and the verb (as in Welsh) but tense affixes are attracted to the verb (as in English). Such a language would have sentences like:

Chris buy-PAST the car.
Will Chris buy the car.

But this combination of properties is not attested in the languages of the world. The space of possibilities is as outlined in Table 6.3. This shows that the verb attraction parameter has priority over the subject placement parameter. A proper understanding of why this relationship holds requires that we adopt a more sophisticated statement of the verb attraction parameter than the one I gave in Chapter 5. The more accurate statement is too technical to be conveniently presented here, however.

As for the null subject parameter, this was originally presented as a matter internal to the Romance languages, distinguishing French (and English) from languages like Italian and Spanish. As such it would be relatively low in the table. Indeed, there are conjectures in the field that only a proper subset of the verb-attracting languages can be null subject languages in the original sense. If these conjectures are correct, then the null subject parameter would be ranked below the head attraction parameter and probably below its partner, the subject placement parameter, as well. That is where I put it. Recall, however, that we saw in Chapter 5 that the grammatical properties of Swahili and other "optional polysynthetic languages" are

TABLE 6.3   Verb Attraction and Subject Placement

	Verb Is Attracted	*Tense Is Attracted*
Subject goes with VP	Welsh, Zapotec	—
Subject goes with AuxP	French, Italian	English, Edo

not too different from those of Spanish and Italian, at least superficially. When the study of null subject phenomena is taken out of its original, Romance-centric context, we may discover that the null subject parameter is essentially the same thing as the optional polysynthesis parameter. This is currently a matter of controversy and ongoing research. While this is being sorted out, I will assume that the null subject parameter is at least partially distinct from the optional polysynthesis parameter, because it makes the table more interesting.

Putting together this combination of hard fact and scientific imagination, I can now present a first draft of a table of languages, as shown in Figure 6.2. This diagram uses the following conventions: If parameter X has logical priority over another parameter Y, then X is written higher than Y and is connected to Y by a downward slanting line. If two parameters are logically independent of each other, then they are written on the same line and are separated by a dash. For convenience, each parameter is assumed to have exactly two possible settings (although this will not necessarily always be true). If there is only one parameter at a level, then it has two branches going down from it, representing its two possible settings. If there are two independent parameters at a level, then there are four branches going out of the dash between them, representing the four possible combinations of settings for those two parameters. (If there were three parameters on a single level, each with two possible settings, then there would be eight branches, and so on.) Since parameter Y is subordinate to another parameter X if and only if Y influences just one of the language types defined by X, it is natural to put Y at the end of the branch that represents the setting of X that Y influences. If there are no further parametric choices to be made given a particular setting of a parameter, then the branch ends in a terminal symbol: *. Beneath this asterisk

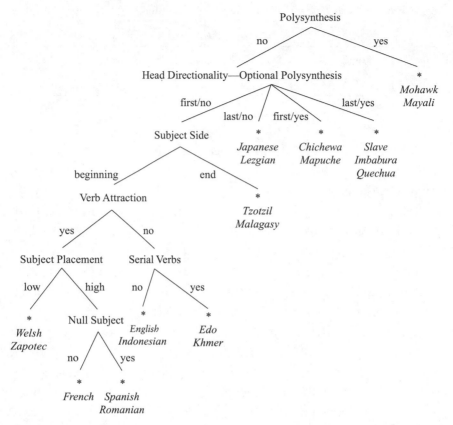

FIGURE 6.2   The Parameter Hierarchy (Preliminary)

I have listed in italics two languages that have this combination of parameter settings. The diagram is complete in that it represents all the parameters and language types we have discussed so far. Structurally similar languages end up being close together in the diagram, whereas structurally very different languages are far apart. For example, English is relatively near to Edo, but it is far away from Mohawk and is almost as far away from Japanese. The table is thus systematic in the same way the periodic table of elements is systematic.

We need a more accurate name for this chart. Calling it the "periodic table of languages" may be evocative, but it is also misleading. For one thing, the diagram lists parameters rather than languages, just as the periodic table lists elements rather than molecules. Second, it is not strictly speaking a table, nor is it periodic. Its name should be more descriptive, if less colorful. Now that we know what it is, I propose to refer to it as the *parameter hierarchy*.

The parameter hierarchy can be compared to another, perhaps more familiar way of charting the relationships among languages, which is by their historical origins. These are usually presented in the form of a family tree, with some languages being in a kind of parent-child relationship (for example, Latin and French), others in a sibling relationship (French and Spanish), still others as distant cousins several times removed (English and Hindi). Figure 6.3 gives a standard representation of the Indo-European family. This mode of representation also parallels the system of biological classification into kingdom, phylum, class, order, family, genus, and species. Both "trees" represent Darwinian relationships of common descent.

Although this kind of family tree representation expresses an undoubted truth, it is not the linguistic answer to chemistry's periodic table. First, it cannot be complete. It has no place for "foundling" languages such as Basque, which are not related to any other known languages. Such languages must of course have ancestors, but they are dead and forgotten, so there is no proper place for them in a family tree. It is true that efforts are under way to relate all known languages back to a hypothetical first common ancestor called Proto-World. But these approaches are highly speculative and controversial, and since critical documentation of languages and their stages is missing, they will probably remain so.

A more serious critique is that the family tree approach is not systematic either. Such a representation shows us in only a crude way what languages are like. It highlights the superficial similarities of vocabulary and idiom and ignores the deeper ways in which languages can differ from their ancestors and siblings. Within Indo-European, for example, Indic languages like Hindi are head-last and thus grammatically more similar to their Dravidian neighbors and to

172

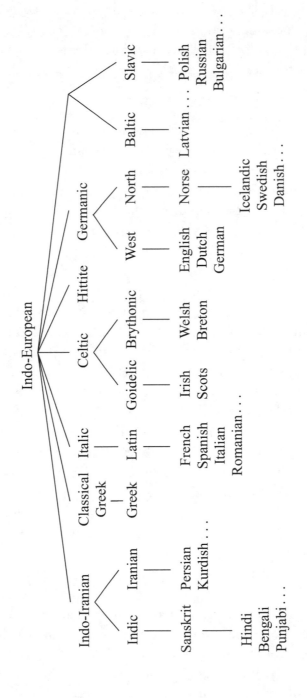

FIGURE 6.3   The Indo-European Family

distant Japanese than to their head-initial relatives like English and Welsh. Similarly, the Australian language Mayali is historically related to the other Australian languages, but in terms of its grammatical structure (and its parameter settings) it is more like faraway Mohawk. These theoretically significant similarities and differences are concealed rather than revealed by standard family trees. For many scientific purposes, we want a presentation that reflects current grammatical properties rather than historical relatedness. Organizing languages historically is a bit like organizing a table of the elements not according to atomic number and valence but according to where their ores can be found. Such a representation could be very useful to mining companies, but a theoretical chemist doesn't care so much where her sample of tungsten came from as what it is and how it works. The same is true for a theoretical linguist. In this respect, the parameter hierarchy is superior to a family tree.

The second distinctive property of the periodic table of the elements is its completeness. Mendeleyev used the systematicity of his table to predict the properties of hitherto unknown elements and thus made the table more complete. The question is whether we can do something similar with the parameter hierarchy. At the moment we can't, at least not to the same degree. It may be that the attempt is simply premature. After all, Mendeleyev knew sixty-one elements at the time when he predicted the existence of three new ones. So far we know of only some eight parameters—not a large base from which to extrapolate. Indeed, it may never be possible for linguists to make predictions exactly like Mendeleyev's. The space of possible parameters is probably not as tightly constrained by the physical properties of the universe as the space of possible elements is. The task of a linguist trying to predict a new parameter might be less like Mendeleyev predicting a new element than like biologists trying to predict the existence of the okapi on the basis of their knowledge of giraffes and antelopes—an impossible task.

This does not mean that the parameter hierarchy is of no value in making predictions. Although we cannot predict the existence of new parameters in any detail (at least not yet), it is well within our grasp to predict the existence of new languages. Such languages should occur whenever the known parameters can combine in a way that has not been observed but that does not lead to a logical contradiction. It is reasonable to conjecture that languages with the theoretically possible combination of properties could be found if we looked for them. We saw one prediction of this type in Chapter 5, when we speculated that there might be such a thing as an object-subject-verb language if Welsh-type verb attraction applied to a language with basic object-verb-subject word order. I have never observed this type of language, but there are claims that Warao and Nadëb might be examples. In a similar way, drawing on our analysis of Welsh and English, we predicted that there would be a language with subjects, verbs, and adverbs positioned as they are in French. Such predictions are more modest than Mendeleyev's: They are not like predicting a new element but only a new chemical compound. Such results are not to be despised, however. That linguists are making such predictions more and more routinely—and turning out to be right—is an important sign of the advancement of the field.

Even a very preliminary parameter hierarchy can stimulate research in other, less rigorous ways. For example, it may have struck you as suspicious that the left side of the hierarchy is much more elaborate than the right side. The "no" branch of the polysynthesis parameter leads to some fourteen further branches that represent different kinds of nonpolysynthetic languages, whereas the "yes" branch is a dead end. Can it really be true that there is only one kind of polysynthetic language? Similarly, the parameter hierarchy includes only two kinds of head-last languages (the Japanese-type and the Slave-type) but eight kinds of head-first languages.

There are good reasons why many of the parameters we have looked at are not relevant to Japanese-style or Mohawk-style languages, but one might presume that there exist other parameters that are relevant to those languages but not to English-style languages. If so, then the overall branching density of the parameter hierarchy would be more uniform. Such thoughts may open our eyes to the possibility of parameters that we have missed, perhaps because of an unconscious English-centric bias. (Indeed, I have written this book with a very conscious English-centric bias, because many of my readers will find it comfortable to use English as a point of reference.)

We would be right to have our eyes opened in this way. It is not surprising that there is more going on in polysynthetic and head-final languages than we have seen so far. In the rest of this chapter, I briefly lay out some possible parameters that are relevant to these languages. Together, they help flesh out the parameter hierarchy and offer a better sense of the overall lay of the linguistic land. (You can pick and choose from the next two sections—or skip them entirely—without harming your ability to comprehend Chapter 7.)

In previous chapters I discussed the various ways that nouns, verbs, and adjectives could combine with other elements to make noun phrases, verb phrases, and adjective phrases. But I took it for granted that all languages have essentially the same distinctions among nouns, verbs, and adjectives that English does. It has been suggested that this is not so. Some languages seem to have more word classes than others. Adjectives appear to be particularly variable, with many languages not having a distinct category of adjectives at all. These languages still have words with meanings like 'white,' 'big,' 'good,' and 'soft,' corresponding to adjectives in English, but these words have the same grammatical behavior as some other word class. In Mohawk and many other Native American languages, the words that English speakers think of as adjectives are actually verbs. One does not use a copular verb like *be* in Mohawk; tense and agreement affixes attach directly to the "adjective" the same way that they do to (other) verbs.

Thikv	kanuhsa'	ka-rakv-hne'.	(Mohawk)
That	house	it-white-PAST	

'That house used to be white.' (literally, 'That house whited.')

Other languages treat words like 'white' or 'tall' as nouns. In these languages, tense and agreement cannot be attached to 'white,' but nounlike affixes can be, including articles and markings for singular or plural. In these languages one may as easily say the equivalent of 'I bought the talls' as 'I bought the chairs.' English, by contrast, forces one to say 'I bought the tall *ones*'; the dummy noun *one* appears along with the adjective in order to carry the plural suffix -*s*. Many Australian languages make adjectives a subclass of nouns in this way:

> Kandiwo   mankuyeng!   (Mayali)
> You/me-give   long
> 'Give me the long one!' (literally, 'Give me long!')

Thus, we seem to have a parameter that can be expressed as follows:

THE ADJECTIVE NEUTRALIZATION PARAMETER

"Adjectives" (words like *white, long, good*) are treated as a
   kind of verb.

or

"Adjectives" (words like *white, long, good*) are treated as a
   kind of noun.

This parameter seems particularly relevant to the nonconfigurational, polysynthetic languages—those that have free word order and do not group words into phrases in the English/Japanese way. There are various reasons why this might be so. For example, the distinctive job of an adjective in languages like English is to combine with nouns to make larger noun phrases, such as *the big white house*. Nonconfigurational languages do not emphasize phrase construction and might thereby have less need for adjectives as a special grammatical category. One could also look at it the other way around: Because these languages do not have adjectives, they are more limited in their ability to construct phrases from individual words. Therefore, they tend to be nonconfigurational or polysynthetic.

D. N. S. Bhat has suggested that the adjective neutralization parameter has repercussions for how the nonconfigurationality of a language is expressed. In all nonconfigurational languages, word order is free, with subjects and objects appearing either before the verb or after the verb. Languages in which "adjectives" are treated like nouns take this freedom a step farther: One part of the subject or object can appear before the verb, whereas another part appears after the verb. Here is an example from Mayali:

Namarngorl   gagarrme   nagimiuk.
Barramundi   he-catch   big
'He's catching a big barramundi [a fish species].'

This kind of "split noun phrase" is common in many Australian languages (see Chapter 2 for a discussion of Warlpiri). It also occurs in some Native American languages such as Sahaptin, and it was found in classical Indo-European languages like Latin and Sanskrit, in which adjectives took the same gender and number endings as nouns. In contrast, split noun phrases are not characteristic of Mohawk and other languages that treat "adjectives" as verbs. So there are (at least) two kinds of nonconfigurational, polysynthetic language, the difference being induced by the adjective neutralization parameter.

Next, let us look more closely at the head-last languages. One property most of these languages share is that their noun phrases have a *case* suffix or particle that indicates the role of the noun phrase in the clause as a whole. Japanese provides a straightforward illustration: The particle -*ga* follows subjects, the particle -*o* follows objects, and the particle -*ni* follows indirect objects:

John-ga   Mary-ni   hon-o   yatta.
John-SU   Mary-IOB   book-DOB   gave
'John gave Mary a book.'

In traditional (Latin-based) terminology, the affix that marks the grammatical subject is called the *nominative* case marker, the affix that marks a direct object is the *accusative* case marker, and the affix that marks an indirect object is the *dative* case marker.

The use of case markers is probably not a distinct parameter in itself but rather an indirect repercussion of the head directionality parameter. Head-final languages almost always have case markers, whereas head-initial languages like English and French typically do not. This was another of Joseph Greenberg's universals (number 41), dating back to the first modern work on typology. Exactly why this correlation holds is not entirely clear, but it is not crucial for us. What is relevant is that languages use (at least) two distinct case-marking systems. The more familiar system treats all subjects the same. If you compare the following intransitive sentence in Japanese to the transitive sentence given above, you can see that the subject is still marked with the nominative affix *-ga*.

John-ga	Kobe-ni	itta.
John-SU	Kobe-to	went
'John went to Kobe.'

This new sentence has no direct object, so it has no noun phrase with the particle *-o*, but otherwise the markings remain the same. Other head-last languages, however, are different. In the following transitive sentence, from Greenlandic Eskimo, the noun phrases carry distinctive case suffixes: *-p* goes on the subject, *-q* on the direct object, and *-nut* on the indirect object:

Juuna-p	atuaga-q	miiqqa-nut	nassiuppaa.
Juuna-SU	book-DOB	child-IOB	send
'Juuna sent a book to the children.'

But in a sentence that does not have a direct object, the subject does not take the *-p* suffix. Rather, it takes *-q*—the ending that otherwise attaches to direct objects:

Atuaga-q   tikissimannngilaq.
Book-(OB?) hasn't-come
'A book hasn't come yet.'

Both Japanese and Greenlandic have systems of case markers that indicate the roles noun phrases play in the sentence as a whole; in
Nichols's terminology (see Chapter 2), both are dependent-marking
languages. But the two systems differ in their details. The Japanese
system is called an *accusative* system, after the Latin name for the
marker of the direct object, and the Greenlandic system is known as
an *ergative* system. An affix like *-p*, which appears only on subjects
of transitive verbs, is known as an ergative case marker. The Australian language Dyirbal and the European language Basque are
other famous ergative languages, whereas Turkish and Quechua are
important accusative languages. There are also some "split ergative"
languages that combine the two types. For example, many present
tense sentences in Hindi use an accusative system, whereas simple
past tense sentences use an ergative system. This, then, is a parameter that divides the head-last languages into subtypes.

Linguists have shown a great deal of interest in the difference between ergative and accusative languages. Much of this interest comes
from the possibility that ergative languages might be a fundamentally
different type from the more familiar accusative languages. Some
people have suggested that the traditional distinction between subject and object (which I have taken for granted throughout this book)
does not apply to languages like Greenlandic. They suggest that the
reason "subjects" do not get a consistent case marker in Greenlandic
is because "subject" is not an important grammatical notion in this
language. Rather, distinct, language-particular grammatical notions
are needed for Greenlandic and other such languages. Other re-
searchers have seen ergative languages as counterexamples to the
verb-object constraint. They say that ergative languages are languages in which the verb combines with the agent before it combines
with the undergoer—the opposite of the rule that we have seen in
most languages. Some have even gone so far as to ascribe deep dif-

ferences in worldview to speakers of ergative languages. They say that speakers of ergative languages have a passive approach to life and therefore focus naturally on the direct object of a transitive verb, whereas speakers of accusative languages have a proactive approach to life and thus focus on the subject.

The bulk of recent research shows that these radical views of ergative languages go too far, exaggerating the differences between the language types. The differences are not entirely trivial: Often which noun phrase in a sentence can most easily be questioned or focused on is different in the two types of languages, for example. But there are also many indications that ergative languages still involve essentially the same notions of subject and object as in other languages. For example, Greenlandic happens to allow a kind of noun incorporation similar to what we saw in Mohawk in Chapter 4. The understood object of the verb can be incorporated in Greenlandic, but the understood subject cannot be:

> Juuna   ilinniartitsisu-siurpuq.
> Juuna   teacher-seek
> 'Juuna is looking for a teacher.'
> *NOT:* 'A teacher is looking for Juuna.'

Examples like this show that Greenlandic distinguishes between subjects and objects in the same way that English, Japanese, and Mohawk do. It even obeys the same verb-object constraint. The difference between ergative languages and accusative languages is localized in the case system and in those aspects of grammar that feed directly off that system. Greenlandic does have subjects and objects, but these notions do not determine case marking in the simplest possible way. I would state the parameter as follows:

THE ERGATIVE CASE PARAMETER

The case marker on all subjects is the same (Japanese, Turkish, Quechua).

or

The case marker on the subject of an intransitive verb is the
same as the case marker on the object of a transitive verb
(Greenlandic, Dyirbal, Basque).

Since this parameter governs case markers, and since case markers are
characteristic of head-final languages rather than head-initial languages,
this parameter is ranked lower than the head directionality parameter.
(But a parallel distinction between ergative and accusative also arises in
agreement systems. This form of ergativity does show up in head-initial
languages such as Tzotzil and Chamorro. Here I leave open what the
connection is—if any—between these two types of "ergativity".)

A second parametric difference that seems particularly relevant to
head-final languages concerns Li and Thompson's distinction between topic-prominent languages and subject-prominent languages.
English is a typical subject-prominent language because the most
basic opposition that clauses are built around is the opposition between the subject and the predicate (roughly, the verb phrase). In
contrast, Japanese is a topic-prominent language: Clauses are normally divided into a sentence-initial noun phrase, marked with a special particle *(wa)*, and a complete sentence, which is understood as a
comment about the initial noun phrase. The *wa*-marked topic phrase
in Japanese can correspond to the subject in English. In this case the
difference between the two languages is relatively slight:

John	wa	sono	hon-o	yonda.
John	TOPIC	that	book-OB	read

'John read the book'; 'Speaking of John, he read the book.'

Yet topic and subject are not always the same thing in Japanese. The
Japanese topic can also correspond to a direct object in English:

Kono	hon	wa	John-ga	yonda.
This	book	TOPIC	John-SU	read

'Speaking of this book, John has read it.'

Sentences like this have both a topic and a subject, showing that the two are distinct. Of course, a kind of topicalization is possible in English as well, as in a sentence like *This book, John has read*. Such sentences are relatively uncommon in English, however, whereas most sentences have a topic in Japanese. The most striking difference between the two systems is illustrated by Japanese sentences like the following:

> Sakana   wa   tai-ga   oisii.
> Fish   TOPIC   red-snapper-SU   is-delicious
> '[Speaking of] fish, red snapper is delicious.'

Here the topic noun phrase does not correspond to anything in the base English sentence; it is an "extra" noun phrase, with no role other than topic. Such sentences are impossible in English but normal in Japanese and other topic-prominent languages. The first two Japanese sentences given above can be seen as variations of this structure: They are special cases in which the topic phrase happens to match the understood subject or object of the comment phrase. But it is not required that there be any such match. This suggests the following form for the parameter:

THE TOPIC-PROMINENT PARAMETER

> A sentence may be made up of an initial noun phrase (the topic) and a complete clause that is understood as a comment on that topic (Japanese).
>
> *or*
>
> No topic phrase distinct from the clause is allowed (English).

The effects of this parameter are actually quite pervasive, showing up in many places other than simple main sentences.

All of Li and Thompson's original topic-prominent languages were from East Asia: Mandarin Chinese, Lahu, Lisu, and Korean as well as Japanese. Languages from other parts of the world that seem to be

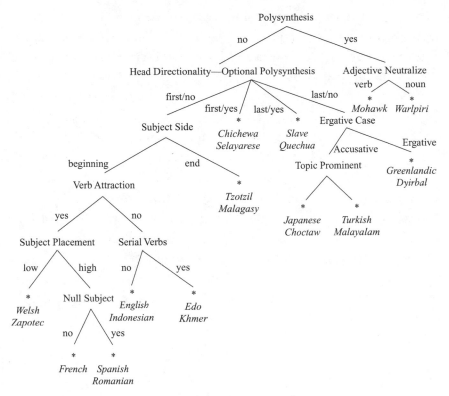

FIGURE 6.4   The Parameter Hierarchy (Revised)

similar include Somali (Africa), Quechua (South America), and Choctaw (North America). Li and Thompson also observe that the topic-prominent languages are a proper subset of the head-final languages (except for Mandarin Chinese, which has mixed word order tendencies)—an observation that has held up well. Thus, the topic-prominent parameter seems logically subordinate to the head directionality parameter. Moreover, all of these languages happen to have accusative rather than ergative case systems. This could be a coincidence. Nevertheless, I put the topic-prominent parameter below the ergative case parameter as well, pending further research into these possibilities.

Integrating this new material, I arrive at my second draft of the parameter hierarchy, shown in Figure 6.4. Although it would be going too far to say that elegant symmetries are emerging, at least the right side of the chart is filling in nicely. And the idea of structuring the "table of languages" around the notion of logical priority seems to be holding up.

———————

There is no reason to expect that all new parameters will be subservient to the head directionality parameter and the polysynthesis parameter. One of the best-studied parameters in recent years concerns whether question words need to move to the front of the clause. In English they do. The corresponding question form of *John bought a new car* is usually not **John bought what?* This order is possible, but only with heavy stress on *what;* in that case the question expresses surprise or horror or asks for the immediately prior sentence to be repeated. The normal form of the question in English is to move *what* to the front of the clause:

What did John buy?

This moving of question words to the front of the clause is not required in Japanese. Rather, question words appear in the same position as other noun phrases in Japanese. For objects, this position is after the subject and before the verb:

John-ga   dare-o   butta   ka?
John-su   who-ob   hit
'Whom did John hit?' (compare: *John-ga Bill-o butta*—'John hit Bill.')

Thus, there is a parameter something like this:

THE QUESTION MOVEMENT PARAMETER

Interrogative phrases must move to the front of the clause
    (English).

or

>   Interrogative phrases appear in the same positions as other
>   noun phrases (Japanese).

There are several other subcases of this parameter and interesting interactions with other grammatical constructions, but I will not go into them here. I do, however, want to raise the question of how this parameter relates to the head directionality parameter. There is a strong correlation between the setting of the head directionality parameter and the setting of the question movement parameter. Head-initial languages *tend* to require the movement of interrogative phrases, whereas head-final languages *tend* not to. But these are not rigid rules or logical consequences. For example, Chichewa is a head-initial language in which question words need not move to the front. The following sentence is a perfectly ordinary question in that language:

> Mu-kufuna   chiyani?
> You-want   what
> 'What do you want?'

Cuzco Quechua, in contrast, is a head-last language in which interrogative phrases need to move to the front. It seems that the question movement parameter is logically independent of the head directionality parameter, although some combinations are clearly more frequent than others. Still, it has been argued that question movement must happen in all polysynthetic languages. (Unmoved question words would act like nonreferential quantifiers such as *nobody,* and these are impossible in polysynthetic languages; see Chapter 4.) Otherwise, general principles of grammar would make it impossible for a speaker of a polysynthetic language to ask questions at all—a limitation that humans would presumably find intolerable (at least my children would). It seems that the proper place for the question movement parameter is below the polysynthesis parameter, at the same level as the head directionality and optional polysynthesis parameters.

Another well-studied area of variation involves pronouns. I mentioned in Chapter 4 that an ordinary object pronoun typically cannot refer to the same person as the subject of the clause. Instead, a special reflexive pronoun must be used. One can therefore observe contrasts like the following in English and many other languages:

John helped him. [ordinary pronoun *him* is not John]
John helped himself. [special reflexive pronoun *himself* is John]

Although most languages share this core distinction, they differ significantly on exactly when the reflexive pronoun can be used. In English a reflexive pronoun can also be used to refer to the direct object of the sentence, as well as to the subject. But a reflexive pronoun that appears in an embedded clause cannot refer to the subject of the main clause; it has to refer to some noun phrase in its own clause. Both of these facts are illustrated in the following complex sentence:

Sue said that Bill revealed Mary to *herself* by a long process of psychotherapy.

In this sentence *herself* refers to Mary (the direct object of the embedded sentence) and cannot refer to Sue (the subject of the main clause). These possibilities reverse if *herself* is replaced by *her*: The ordinary pronoun in this position could mean Sue but not Mary.

Reflexive pronouns in other languages have somewhat different opportunities and limitations. For example, the Chinese reflexive *ziji* can refer to the subject of its own clause, as in English. Unlike English, however, it can also refer to the subject of the larger sentence. But *ziji* can never refer to a direct object. The possible meanings of the following Chinese sentences are thus different from those of their English equivalents.

Zhangsan   renwei Lisi   hai-le   ziji.
Zhangsan   think   Lisi   hurt   self
'Zhangsan thought that Lisi hurt (him)self.'

(The hurt person could be Lisi or Zhangsan.)

Zhangsan	gaosu	Lisi	ziji	de	fenshu.
Zhangsan	tell	Lisi	self	's	grade

'Zhangsan told Lisi his own grade.' (It must be Zhangsan's grade, not Lisi's.)

To account for differences like these, linguists have proposed two parameters:

THE REFLEXIVE DOMAIN PARAMETER

A reflexive pronoun must refer to the same thing as some other noun phrase that is contained in the same clause (English).

or

A reflexive pronoun may refer to the same thing as a noun phrase outside its clause (Chinese).

THE REFLEXIVE ANTECEDENT PARAMETER

A reflexive pronoun must refer to the same thing as a subject noun phrase (Chinese).

or

A reflexive pronoun can refer to any noun phrase that comes before it in the clause (English).

These parameters also seem to be largely independent of the others we have studied. Thus, the word order of Chinese and English is roughly the same, even though their reflexive pronouns work differently. Japanese, in contrast, has a very different word order from Chinese, but its reflexive pronouns work the same way. For the most part, it seems that any type of reflexive pronoun can appear in a language with any type of word order. (An exception is the polysynthetic languages, which because of their unique structure cannot have certain types of reflexive pronouns at all; see Chapter 4.)

A closer look at the case of reflexive pronouns brings us to the limits of the parametric approach to linguistic diversity. Throughout this book, I have assumed that language can be divided into at least two facets: the lexicon (the stock of words of a language) and the grammar (the recipe for putting words together into phrases and sentences). Parameters are a way of thinking about differences among languages' grammars. The obvious differences among languages' lexicons I have largely taken for granted. But here we see a place where the two topics meet. The difference between Chinese and English, which I have presented as a parametric difference, could just as easily be seen as a difference in the two languages' lexicons. We know that sometimes one language will have a word for something and another language will not. For example, the languages of Africa and the Americas generally had no word for *book* prior to contact with Arabic or a European language, for the somewhat dull reason that the speakers of these languages had no exposure to books. (One can usually tell whether Muslims or Christians got to a certain region of Africa first based on whether libraries contain *buku* or *kitabu*.) Europeans, conversely, had no words for *chocolate* or *coyote* prior to contact with the Aztecs. Such words can easily be added to or subtracted from a language, with little effect on the language as a whole.

Sometimes, however, an entire class of words may drop in or out of a language, particularly if it is a small class to begin with. For example, we have seen that many languages, including Japanese, have no words comparable to the articles in English. Although this is primarily a lexical difference, it has syntactic effects, at least to the extent that "article phrases" are formed in English but not in Japanese. The syntactic effects will be greater if it turns out that there are special grammatical functions that can be fulfilled only by an article phrase and not by its close cousin the noun phrase. Current thinking suggests that there is actually quite a bit of linguistic variation of this type. For example, some languages do not make tense distinctions among past, present, and future the way that English and other Indo-

European languages do. The Hopi language, analyzed by Benjamin Whorf in the first half of the twentieth century, is a famous example; others include Eskimo and Mohawk. These languages zero in on somewhat different distinctions, or they add time-denoting adverbs like 'now' or 'later' or 'yesterday' to an otherwise ambiguous sentence. Such languages can be seen as lacking tense auxiliaries in their lexicon. As a result, they do not form auxiliary phrases, and thus their grammar differs from that of other languages.

Given that this kind of lexical variation is possible, the question arises whether some of the differences among languages that we have treated as parameters should actually be treated as differences in the lexicon. This is a center of current debate, and for the just-discussed differences in reflexive pronouns the lexical solution looks like the correct answer. The hint that this might be so is that both the reflexive domain parameter and the reflexive antecedent parameter refer explicitly to a very small class of words: the reflexive pronouns. In English this class consists of only a handful of closely related items: *myself, yourself, himself, herself, itself, ourselves, yourselves, themselves,* and *each other.* Since this class is small, it is easy to imagine that some language might have none of these elements. Instead of saying that there are two types of *languages* that differ in how their reflexive pronouns work, we can say that there are two different types of *reflexive pronouns* that a language may have (let's call them "type A" and "type B"). If there are no parameters that refer directly to the behavior of reflexive pronouns, then type A pronouns will behave the same way in all languages, as will type B pronouns, although the two types will behave differently from each other. Two languages could look syntactically different if one happened to have only reflexive pronouns of type A and the other had only type B. There would be no difference in parameters, but there would be a lexical difference with grammatical consequences.

This reasoning applies well to the putative difference between English and Chinese with respect to reflexives. In addition to *ziji*, Chinese has a second reflexive pronoun, *taziji*. This pronoun behaves more like *himself* in English. For example, it can refer to the subject of its own clause but not to the subject of the main clause (contrast this with the example using *ziji* earlier in this chapter):

Zhangsan   renwei Lisi   hai-le   taziji.
Zhangsan   think Lisi   hurt   himself
'Zhangsan thought that Lisi hurt himself.'
(The hurt person must be Lisi, not Zhangsan.)

Taziji also resembles *himself* in that it is a compound form, made up of an ordinary pronoun *ta* 'him,' plus a reflexive element *ziji* 'self.' This shows that the parametric approach, which would treat all reflexive pronouns the same, is wrong for this case. It is not an overall property of the Chinese language that reflexive pronouns work in a particular way. Rather, it is a property of the simple reflexive pronoun *ziji* in Chinese that it works in a certain way. English happens not to have a simple reflexive comparable to *ziji*, even though all the parameters are set in the same way; in English it is impossible to say **Chris hurt self*. Thus, I refrain from adding the reflexive parameters to our official list. More generally, the bottom fringe of the parameter hierarchy shades off into these kinds of lexical issues.

Should any of our other parameters be recast as lexical differences in this way? The natural candidates for such recasting would be any parameters that mention a particular class of words, particularly if it is a small, closed class. The head directionality parameter is not vulnerable to this kind of reanalysis, nor is the polysynthesis parameter. They are too general to be tied to a single type of word. But the question movement parameter might be a candidate, since it mentions interrogative phrases, which are formed out of a small number of designated question words—in English *who, what, which, why, where, when,* and *how.*

The clincher for a lexical reformulation will always be to find cases, as we did with Chinese, in which both of the behaviors described by a putative parameter are found in the same language, depending on what word is used. For example, we would know that the question movement parameter was a matter of lexical differences if we found languages in which some question words always moved to the front of the clause (as in English) and other question words did not (as in Japanese). This could be a language, for example, where one asked *Whom did you see?* but *You saw what?* As far as I know, this situation is not found in any language. Similarly, we would know that the head directionality parameter was really a lexical matter if we found languages in which some verbs always came before their objects and other verbs always came after them. This, too, seems never to happen. All tenses attract verbs in French, whereas all verbs attract tenses in English; this vindicates the verb attraction parameter. Overall, these seem to be properties of languages themselves rather than of particular words in the languages. It seems correct to keep thinking of all or most of the material we have discussed in terms of parameters, not as side effects of the lexicon. Although the exact location of the parameter hierarchy's bottom fringe might be questioned, the existence of the hierarchy itself seems beyond doubt.

Mendeleyev's first periodic table had some notable imperfections (see Figure 2.2). It was oblivious to a whole class of elements, the noble gases, whose existence he never dreamed of. Mendeleyev predicted the discovery of germanium and gallium, but not neon and argon, even though there is plenty of argon all around us. The first periodic table was also organized around atomic weight rather than atomic number. This is close to the same thing but leads to a few needless irregularities. I have no doubt that my precursor to a table of languages has many similar flaws. There are probably things that should be included that are undreamt of in my philosophy. I would also not be surprised if the organizing factor I have chosen turns out to be close but not quite right. Nevertheless, the flaws of Mendeleyev's table did not prevent it from being an invaluable landmark of chemistry, completing one phase of research and opening the doors to another. With that in mind, I close this chapter with some additional thoughts about the potential importance of a parameter hierarchy along the lines I have been imagining.

I doubt the parameter hierarchy will ever achieve the kind of semi-mathematical rigor of Mendeleyev's table, given that linguistics is far removed from the symmetries of fundamental physics. But it will show which languages are grammatically more similar and which are more different, independent of any shared history. The structure of the parameter hierarchy does not tell us how people use the parameters while speaking or understanding a language, but its logical structure could be a very useful guide to linguists who are beginning to analyze a new language. For such linguists, it would make sense to begin at the top of the hierarchy and work their way down, deciding first whether a language is polysynthetic or not, then whether it is head-first or head-last, and only then considering whether it has serial verb constructions. There is no rush to look for a serial verb construction if the language being studied belongs to a type that is known not to permit them. In fact, this is roughly what linguists already do. A parameter hierarchy would permit them to do it more consciously and systematically.

Even more exciting is the possibility that the parameter hierarchy might provide a framework for understanding how children manage to learn something as complex as a natural language with little or no explicit coaching. The problem children face is essentially the same as that of the practicing linguist: They are trying to deduce the recipe for the language being spoken around them so that they will be able to make sentences of their own. It would make sense if children, too, instinctively work their way down the hierarchy, taking advantage of its logical structure to avoid agonizing over needless decisions.

To evaluate this possibility, we can consider what is involved in learning the grammar of French as opposed to grossly similar languages like Spanish, Welsh, or English. This is where the parameter hierarchy as we know it is the deepest: The settings for six parameters must be determined. The question is whether those settings are acquired spontaneously in the order that the parameter hierarchy predicts.

Unfortunately, nothing is known about the acquisition of two of the first three parameters. The polysynthesis parameter and the subject side parameter have not been studied from this perspective. This is primarily for practical reasons: Languages like Mohawk and Malagasy, for which the values of these parameters are crucial, are relatively small and remote languages, and few children are learning them.

These problems do not beset the head directionality parameter, however. The earliest rigorous studies of language acquisition showed that children learning different languages show systematic differences in word order. Indeed, the very first two-word utterances of children—which typically appear between eighteen and twenty-two months—show that they have already set the head directionality parameter. An English-speaking one-year-old's first sentence might well be something like *Give cookie!* Japanese one-year-olds have comparable interests, but their grammar is already different: They begin with *Cookie give!* The prediction that this parameter would be one of the first ones set is thus correct.

After the major word order parameters have been established, the next one we expect children to attend to is the verb attraction parameter. Vivian Déprez and Amy Pierce have shown that this, too, is learned quite early. They studied the order of subjects, verbs, and negative adverbs in children learning French and English, starting at around twenty-one months. They found that children at this time already showed evidence of knowing that verbs move to tense in French but not in English. English-speaking children regularly utter sentences like *Not have coffee,* with the verb coming after the negative adverb *not.* French children at the same stage say *Marche pas* ('works not,' meaning 'it doesn't work') and *Veux pas lolo* ('want not milk'). In these examples the verb has moved to tense and thus comes before the negative adverb *pas.* French children also use infinitive verb forms with negation. Since these verbs have no tense value, we predict that they do not move to tense and hence should not come before *pas.* It is significant that when French children use an infinitive the verb consistently comes after *pas.* Examples are: *Pas casser* ('not to-break') and *Pas attraper papillon* ('not to-catch butterfly'). French-speaking children of twenty-two months thus have worked out that tensed verbs move forward but infinitival verbs do not.

According to the parameter hierarchy, the next matter to consider should be the subject placement parameter, whether subjects are attached to auxiliary phrases or to verb phrases. Déprez and Pierce show that this is not decided until slightly later, at about twenty-four months. Both French and English children go through a stage in which their verbs are properly placed but their subjects are not. Thus, on their second birthday English children are wont to say:

No I see truck.
No Leila have a turn.

Here the subject comes after the negative marker, not before it as in adult English. (These examples also prove that children don't just mindlessly mimic the orders they hear adults use.) French children of the same age use sentences in which the subject follows the tensed verb:

Tombe Victor.
Falls Victor   ('Victor falls.')
Veut encore Adrien du pain.
Wants more Adrien bread ('Adrien wants more bread.')

This can result in verb-subject-object orders, just as if the young French children were speaking Welsh. At this point the first four parameters are set correctly but the last two are not. English-speaking children begin to put the subject at the front of the sentence, attached to an auxiliary phrase, at an average age of 24.5 months. That, then, is roughly when they determine the correct setting for the subject placement parameter.

At the bottom of the left side of the parameter hierarchy is the null subject parameter, which determines whether clauses are required to have subjects or not. This is correctly predicted to be learned relatively late. It has long been known that children learning English or French omit subjects as much as 50 percent of the time. They are much less likely to leave out direct objects. This results in utterances like the following:

Want go get it.

Not making muffins.

Similarly, French children of twenty-six months say *Est pas mort* ('Is not dead') for *Il n'est pas mort* ('It is not dead'). Nina Hyams has observed that such sentences are grammatical in languages like Spanish and Italian. She concludes that these children have not yet determined the setting of the null subject parameter. The average age for establishing this parameter value was twenty-seven months in the children she studied—somewhat later than the setting of the subject placement parameter, as predicted.

Less is known about the stages of learning for languages on the right side of the parameter hierarchy. Clifton Pye's work, however, provides one encouraging result. He studied the acquisition of ergative systems as opposed to accusative case systems, a parameter relevant to head-final languages. He reports that Turkish and Japanese children begin to master their languages' accusative case system at about twenty-four months. This is significantly after their first signs of learning that those languages are head-final. Similarly, children acquiring a Mayan language make few mistakes with that language's ergative marking after the age of two. This is the order of acquisition we would expect, given that the ergative case parameter is subordinate to the head directionality parameter.

Overall these results are encouraging for the view that the parameter hierarchy provides a logical flowchart that children use in the process of language acquisition. There are still many parameters that need to be studied in this way. Also, methodologies need to be developed for timing these stages of acquisition accurately, since the pace of acquisition varies considerably from child to child, and many changes come very rapidly between twenty-two and thirty months. If the pattern holds up under further scrutiny, however, this will be a strong source of evidence in favor of the parameter hierarchy.

Another long-term benefit of the periodic table of the elements was that it prepared the way for an understanding of quantum mechanics and deeper theories of the atom. To the casual observer, the periodic table of the elements has a somewhat odd structure (see Figure 6.1). It is not a simple table with a uniform number of cells in each row, the way that word processors like tables to be. Rather, the number of cells per row increases systematically as one goes down the table. Every second row is longer than the row preceding it by an ever-increasing amount: The first row has two members, the second row adds six more cells, the fourth row adds ten more, and the sixth row adds fourteen. There was no good reason why this should be so in Mendeleyev's day; it just worked. We can now see that this quirky structure reveals deep quantum mechanical principles—in this case, about how electrons fill orbitals in the atom. Some chemists have even proposed high-tech, three-dimensional versions of the periodic table in an effort to capture these quantum mechanical regularities more perspicuously (see Figure 6.5). Similarly, there is little doubt that the parameter hierarchy will have some quirky features, with less than uniform branching density and imbalances in its structure. We can hope that these imbalances will provide important clues to the deeper properties of human language and cognition. They could perhaps lead to something like a quantum mechanics of the human mind. As we approach the day when we can present a true "periodic table of languages," we look forward to the fulfillment of one historic phase of human inquiry, the effort to characterize the nature of the differences among human language. But more than that, we can look forward to making a new and deeper set of questions open to meaningful inquiry for the first time.

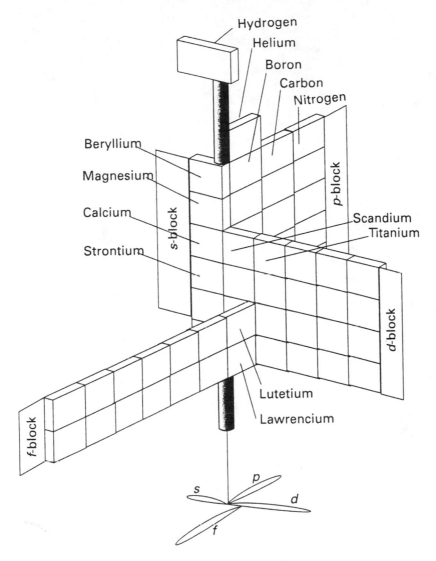

FIGURE 6.5   A Quantum Mechanical Periodic Table

7

Why Parameters?

Suppose the previous six chapters have convinced you that there might be something to the parametric theory of linguistic diversity. Suppose that by now you have some sense of what a parameter is and how parameters can reconcile our conflicting senses of the sameness and difference of languages. Suppose the elegance of the overall conception or the details of word order in Tzotzil has elicited from you a response of "Maybe so." Then it is appropriate, in this last chapter, for us to sit back and speculate. *Why* should it be like this? Why should languages be so similar in their recipes and yet so different in their observed forms? Why should so much of grammar be predetermined, yet not all of it? Why should some of the undetermined parts be located in the central core of the recipe of phrase formation, precisely where they have the greatest effect on sentences? If engineers had consciously wanted to create a system that maximized the apparent differences among languages while minimizing their actual substantive differences, they could hardly have found a better way than by using parameters. But why would this be desirable, and how could it have been accomplished by social or biological evolution?

In short, why are there parameters?

Much of what I have to offer by way of speculation on these grand questions is negative. In particular I wish to resist a quick reduction of the parametric theory to either of the current intellectual world's broad explanatory paradigms: the political dynamics of cultural transmission and the survival dynamics of evolutionary biology. Both of these approaches to explanation have a certain initial plausibility, especially given that we are already immersed in them. But neither seems adept at accounting for the fine structure of linguistic diversity. This diversity always comes out as something extra, mentioned as an awkward footnote if at all.

Still, my speculations will not be purely negative. By taking a brief look at the factors at work in the actual historical process of language diversification, we can, I believe, glean hints of what parameters are related to. If we can find out what parameters interact with, then we will be in a better position to guess where they fit into the world. In fact, they seem to form a piece with some of the notorious unsolved problems of cognitive science, including our sense of free will, intentionality (meaning), and our possession of a priori knowledge. I cannot offer a bold new explanatory paradigm that accounts for these mysteries and includes parameters as a special case. (If I could, I would have announced it much earlier in this book—such as on the front cover.) If one must be left with a mystery, however, it is of some interest to know what kind of mystery it is and how it relates to other mysteries. We may have not caught our killer yet, but at least we have some evidence and an idea of what file to record it in.

———————

Probably the most widely held view outside of formal linguistics is that languages differ for the same reasons that other aspects of culture differ. They differ because the local traditions of an ethnic group are passed on to children through learning and acculturation. This perspective dominates the humanities and social sciences, including anthropology, from whose roots modern linguistics has partly grown. These disciplines tend to think of human nature as being relatively plastic, and the molding influence of culture on the individual is seen as an irresistible force. They also tend to emphasize the arbitrary and conventional aspects of language, such as nuances of particular words and the structure of culturally related terminology concerning matters like kinship or religion. From this perspective, it is no surprise that languages are different. What is unexpected is that there are limits to the differences and that many of the differences can be characterized in a precise, quasi-mathematical way.

Knowledge of a language is undeniably part of one's cultural heritage. A good portion of one's identity as a member of a cultural group comes from being able to speak the group's language. Much of our cultural knowledge is expressed to us in that language. Also, there is no doubt that aspects of language are learned in the same context as other aspects of culture, as children interact with people around them. Thus, it might seem artificial to distinguish language from other kinds of cultural transmission.

Yet there are real limits to how much of language variation can be understood in this way. In particular, the grammar of a group's language does not seem to be correlated with other identifiable features of their culture. As we look at how the different language types are distributed around the world, there is no hint of a significant interaction between language type and cultural type. Japanese, Mongolian, Malayalam, Turkish, Basque, Amharic, Greenlandic Eskimo, Siouan, Choctaw, Diegeno, Quechua, and New Guinean languages are all head-final languages. Chinese, Thai, Indonesian, Arabic, Russian, French, Yoruba, Swahili, Salish, Zapotec, Mayan, and Waurá are all head-initial languages. Is there anything in the pattern of cultural interactions or the basic worldview of the first group of people that consistently distinguishes them from the second group? Is there any causal relationship between how they order their words and how they experience life? So far as anyone knows, the answer is no. The same seems to be true for the other parameters. The culture of a group is typically more similar to that of its neighbors than to the culture of people who speak grammatically similar but historically unrelated languages elsewhere in the world. For example, the cultures of the various native groups in California were very similar despite the great linguistic diversity in that area before the arrival of Europeans. In Australia the predominant cultural features of the northern tribes were much like those of the southern tribes, even though the northern tribes spoke polysynthetic languages and the southern tribes did not. Furthermore, ethnic groups seem able to change their language without changing their culture, and vice versa. Many features of Mohawk society have changed radically over the past 350 years. Nevertheless, Mohawk Catholic nuns and high-rise construction workers use the same syntax as their predecessors who burned tobacco to the spirits and subsisted by growing squash, pole beans, and corn. Conversely, some Mohawks now speak their traditional language and others do not; nevertheless, they are members of the same culture. Thus, although language is culturally transmitted, it seems to be independent of other aspects of culture and not explainable in terms of them.

The other weakness of the standard anthropological perspective is that it has little to say about the limits on linguistic variation or the deep regularities that are found in this variation. Edward Sapir, the greatest anthropological linguist of the first half of the twentieth century, claimed explicitly that "language is a human activity that varies without assignable limit," and many of his contemporaries said similar things. But since Joseph Greenberg (himself an anthropologist) presented his first work on language universals (see Chapter 2), it has become increasingly clear that this is not the case. There are many ways of mixing word orders or marking grammatical relationships that are perfectly imaginable but are not found in languages of the world. Nor can these gaps be attributed to limitations in the acculturation process. If children are really blank slates, ready for culture to write upon, why should it be any harder to transmit to them an object-verb-subject language than a subject-verb-object language? If the mighty kindergarten can teach my children to pick up their toys and wash their hands, why can't it do this? The asymmetries in word order that we observe seem to show that children have strong predispositions in favor of speaking a subject-verb-object language and against speaking an object-verb-subject language (perhaps even stronger than their predisposition not to pick up their toys).

Even where there is variation across languages, it is often more regular and systematic than one would expect of a cultural product.

Mohawk is a polysynthetic language, although the Siouan languages it is probably related to are not. Is it plausible to think that at some point the Mohawk elders grew weary of their macro-Siouan heritage and decided at a tribal council that from then on they would mark all the participants of an event on the verb that expressed that event? There are many reasons why this scenario is implausible. Ordinary speakers of a language who are not trained grammarians are seldom aware of the constituent parts of their language or how those parts are combined. For example, even well-educated speakers of Mohawk do not stop to realize that in a word like *wa'khninu'* ('I bought it') it is the *k* sound that expresses the subject 'I.' If people are not aware of the parts of their language, how can they plan them? And how could they plan them systematically, without missing a relevant case?

Even if the Mohawk elders had wanted to change their language in this way, it is not clear that they would have had the clout to do so. Even now, with the bureaucratic power of modern states, language-planning projects have a very limited record of success. The eminent French Academy has not stopped Parisians from adopting anglicisms like *le weekend*. Why should we think that loosely knit traditional societies would be more successful? Suppose instead that the change from Proto-Macro-Siouan to Mohawk was not consciously planned by anyone, as seems likely. Then why was it so systematic and complete? Again, it seems more plausible to think that parameters reflect natural regularities of the human mind, born into all of us, rather than the results of cultural decision, whether conscious or unconscious.

Here we can add a word about the underpinnings of postmodernism as an influential contemporary viewpoint. I hasten to note that I am no expert on postmodernism and am not qualified to analyze the movement as a whole. One does not have to be an expert, however, to know that it developed in part out of some of the mid-twentieth-century trends in anthropology and linguistics that we have been discussing. The postmodern perspective is thus rooted in certain assumptions about language that I can comment on. Jacques Derrida in particular developed his very influential views in reaction to the Swiss structuralist linguist Ferdinand de Saussure. A contemporary of Boaz, Saussure is famous (among other things) for emphasizing the arbitrary relationship between the sound of a word and its meaning. In itself there is nothing so deep or remarkable about this observation: He was simply pointing out that there was nothing special about the sounds *d, o,* and *g* that makes their concatenation particularly suited to referring to domestic canines. On the contrary, the sounds *chien* (French) or *perro* (Spanish) or *erhar* (Mohawk) or *ekita* (Edo) denote man's best friend just as effectively. What is crucial is that there be a conventional association between sound and meaning that is shared by a group of speakers. Yet Derrida attaches great metaphoric significance to this arbitrariness and conventionality at the roots of language. He also points out (correctly) that the exact meaning associated with a sign like *dog* is hard to pin down. This, too, is no great surprise to anyone who has been asked to define a word; it is a tricky task, and no one ever gets it completely right. There is a fluidity to how words are used that greatly complicates things. On a particular occasion, a speaker might refer to something as a dog in a loose or metaphorical sense that would not carry over to other times or uses. Now, if the most basic pieces of language are arbitrary, fluid, and not directly related to meaning, it is easy to conclude that all of language is like this. This leads naturally to the bewildering and ever-shifting postmodernist view of the world, in which nothing has a lasting or general meaning.

Given the linguistic research presented here, this postmodern line of reasoning does not really go through. The error is simply the common one of seeing words as the (only) atoms of language. Many of Derrida's observations about language are valid at the level of words. For this reason, the area of lexical semantics is perhaps the most problematic and least-developed area of contemporary linguistics. But words are *not* all there is to language; there is also grammar, the principles for combining words into sentences. One might think that if words are arbitrary and fluid in their meanings, larger constructions built out of them would be even more arbitrary and fluid because the indeterminacies of meaning would compound and magnify each other. Yet the opposite is true. Sentence structures are more rigidly and universally specified than word meanings are, as determined by the basic recipe for language together with its atomic variations, the parameters. Thus, whereas the lexicons of different languages vary widely, their grammars do not.

The associations between form and meaning at the level of phrases and sentences also tend to be universal and not arbitrary. For example, we have seen that there is a broad tendency across languages for words that are related semantically to form phrases, so that they end up next to each other. Thus, even languages as different in their syntax as English and Japanese have essentially the same patterns of grouping into phrases. There are also universal rules that say that the agent of an action is expressed as the subject of the verb and the undergoer of the action is expressed as the object of the verb. Furthermore, the object of the verb consistently forms a phrase with the verb that does not include the subject (the verb-object constraint). Perhaps these relationships between form and meaning are ultimately arbitrary and conventional, but they are not variable. The entire human race seems to be fixed on the same conventions. Paradoxically, then, sentence-level form and meaning is much better understood than word-level meaning. Moreover, these constraints on sentence structure are powerful enough that they help narrow down the meanings of the words in the sentences. Although the meaning of a word in isolation is shifty and problematic, the same word in the context of a sentence is usually much less problematic. In context the word's meaning is constrained by both the sentence structure as a whole and the meaning of the other words around it. We can get away with flexibility in word meaning precisely because we know that the linguistic context will clarify what we intend (except, perhaps, in continental philosophy texts). The severe difficulties in dictionary writing are not an accurate measure of the difficulty of interpreting a sentence. Thus, the results of current linguistic research tend to refute the assumptions about language that lie at the foundations of the postmodernist perspective.

Let us turn to contemporary culture's other great answerer of why questions, evolutionary biology. How can we assess its prospects for explaining the existence of parameters?

From an evolutionary perspective, it is not surprising that all humans would share a sophisticated inborn propensity to do something complex. Human language can be thought of as being similar in kind to the many highly sophisticated abilities that characterize other members of the animal kingdom: echolocation in bats, web spinning in spiders, navigation in migratory birds, and so on. Thus, the innate and commensurable aspects of language are more or less expected. Moreover, the obvious usefulness of our language abilities invites an evolutionary approach. It is easy to see that language has contributed mightily to our survival, reproduction, and general remaking of the world in our own image. Human language makes possible the representation and transfer of detailed knowledge about the world, including knowledge that is useful to survival—such as how to prepare bitter manioc so that it is edible rather than poisonous. Language also makes possible sophisticated cooperative efforts, such as mastodon hunting. It is no mystery that a species with an innate ability for complex propositional language can thrive in comparison to competitors that lack such abilities. The theory faces the usual puzzles, of course, such as how physical DNA sequences in the human genome could induce as abstract a thing as our innate language recipes. People have no more idea about how this works in practice than they do about any other complex species-specific ability. There is also the puzzle that human language is unique to our species, with no clear analog in even our closest relatives. Together with the fact that language abilities leave no direct traces in the fossil record, this makes it very difficult to say exactly how language evolved and what its less-developed stages were. As Steven Pinker points out, however, that only one surviving species has anything much like human language is no more of an embarrassment to evolutionary theory than that only one surviving species has a proboscis as versatile and sophisticated as an elephant's trunk. Thus, undaunted by the paucity of hard evidence (and in marked contrast to their earlier reticence), linguists and related academics have recently made a flurry of proposals concerning the evolution of language, some of which have been presented to the public.

But in light of the theme of this book, language is the easy part of the story. Like the writer of Genesis, we don't need a story about the origins of human language; we can take it for granted as an intrinsic part of the general story of the origin of humans themselves. What we need is a story about why languages differ. The real mystery for a biological theory of linguistic diversity is not why there should be an innate recipe for language common to all humans but why that recipe should include parameters. Why doesn't our innate endowment go the whole way and fix all the details of the grammar of human language? One might say that the genome (whose expressive capacity is turning out to be more limited than we had thought) cannot waste its resources on picky details at the edges of language. But it is not only at the edges of language that we find parameters; we also find them at the core. Why should an innate recipe for language allow that kind of variation? Amid the cacophony of recent writing about the evolution of language, the silence on this point is striking. Rarely is the question even framed. Philip Lieberman and Andrew Carstairs-McCarthy say nothing explicit about it. Neither does Derek Bickerton, except for a fleeting reference in one of his early books, which appeals to cultural influence in the way we have already rejected. The linguist Frederick Newmeyer faces one specific question of this type in passing but immediately confesses he has nothing to say about it. One would think that linguistic diversity was an obscure or trivial matter rather than a major political issue in many parts of the world.

The question can be framed somewhat more sharply in terms of the standard thought experiment of evolutionary theory, following the lines of a recent mathematical study by Martin Nowak, Natalia Komarova, and Partha Niyogi. Simplifying their assumptions somewhat, let us suppose that there were three kinds of humans: *Homo rigidus*, whose genome specified a completely fixed grammar; *Homo whateverus*, whose genome did not specify any principles of grammar at all; and *Homo parametrus,* with an intermediate genome that specified many fixed principles but also left some options open, to be set by experience. Assuming that all three compete for the same resources, which type of individual would tend to leave the largest number of offspring?

We can rule out *Homo whateverus* immediately because the kind of language that is sufficient for teaching proper manioc preparation and organizing successful mastodon hunts is presumably too complex for children to learn reliably without a head start. Even if such a language is not too complex to be learned in absolute terms, all that is needed is for the children of *Homo rigidus* and *Homo parametrus* to acquire it more quickly, more accurately, and under a wider range of situations, and *Homo whateverus* will go extinct. In the ongoing battle between the children of *Homo rigidus* and the children of *Homo parametrus,* however, it is not at all clear who would gain the upper hand, since (by hypothesis) both succeed in reliably acquiring a language with high representational and expressive capacity.

At first it might seem easy to give the nod to *Homo parametrus* because we know that in many other cases a combination of innate endowment and learning from the environment turns out to be optimal. Bees, for example, have a complex cognitive system that allows them to find their way home from flower fields by calculating the changing angles of the sun. It is clear that the overall outlines and many of the details of this sophisticated navigational system are hardwired into the bees; they do not learn the relevant trigonometry the way we do in high school. But the system also has a few "parameters" that the bee's genome leaves open to be set by experience. There is an obvious practical reason for this: The angles of the sun are different at different latitudes. If the specific angles were hardwired into the bees, that particular lineage could function only at one latitude. By leaving this factor open, bee species can do well at any latitude, needing only to do some preliminary observations of the sun as they get started. Similarly, bees have some innate ability to recognize potential flowers, and blackbirds have some innate ability to recognize potential nest-robbing birds, but both also do some learning to attune themselves to the precise opportunities and dangers of their particular surroundings. Given the prevalence of such cases in the animal kingdom, some writers are quick to conclude that a mixture of innate knowledge and learning from the environment is the optimal strategy for language, too. Nowak's math confirms that a language recipe that allows a choice of languages is evolutionarily favored as long as the extra languages offer "new advantageous grammatical concepts" that enable people to communicate about new kinds of events in the environment. And it is a striking fact about human beings that we have found ways to live and reproduce in every ecosystem, thanks in large part to the knowledge that can be represented and communicated in our languages.

On closer inspection, however, the analogy between human language and bee navigation does not apply at all. It is true that much of the greatness of the human language capacity comes from its being flexible enough to cope with new situations. But the flexibility to acquire different grammatical systems does not contribute significantly to this virtue. The ability to discuss an infinite number of contingencies seems to be present to an equal degree in every human language. Surviving and reproducing under new conditions might require coining some new words, and it certainly requires forming new sentences. But creating new words and sentences is routine in any language and does not bear on the dimension of flexibility that parameters make possible. The ability to put heads last in phrases rather than first did not make it possible for people to better discuss new hazards or pleasures when they colonized the Japanese islands. Neither does the polysynthetic character of Mohawk make it particularly suited to life in a temperate forest. There is no plausible link between the grammatical properties involved in parameters and the ecological characteristics of environments. Grammatically similar polysynthetic languages are found in the Arctic tundras of Camchatka, the deserts of the American Southwest, and the tropics of northern Australia. Head-final languages are spoken on the steppes of central Asia, in the mountainous Andes of South America, in the jungles of New Guinea, and many other places. For that matter, native speakers of English itself are now found in all ecosystems and climates. Their vocabularies may have adjusted somewhat as a result (although even these adjustments are quite modest), but there has been no pressure for their parameter settings to change along with their physical surroundings. Atlas publishers will never be able to save money by having their climate maps do double duty as language maps. The one language of a *Homo rigidus* could (given suitable vocabulary) have exactly the same expressive power as any of the languages learnable by *Homo parametrus*. And when there is no inherent difference in the biological value of the potential languages, Nowak's calculations show that there is selection pressure to reduce the number of languages that can be learned—with the one language of *Homo rigidus* being the limit. Not only is there no environmental explanation for grammatical diversity, but the best mathematical simulation actually argues against it. Yet the diversity exists.

There is only one important feature of our environment that we relate to differently depending on the grammatical details of the language we speak: other people. Sometimes it is fun and even useful to speak a language that your people can understand but that listeners-in cannot, as the Navajo Code Talkers demonstrated vividly. More generally, we tend to view people who talk like us as potential allies, relatives, and mates, whereas people who do not talk like us are potential adversaries (or mates). There are possible advantages to dividing humanity up into smaller teams, and language differences seem like an effective way to do this. Thus, people like the physicist Freeman Dyson have proposed that linguistic diversity "was nature's way to make it possible for us to evolve rapidly" by creating isolated ethnic groups that can evolve biologically and culturally in different ways, thus diversifying our survival portfolio as a species. But even if this diversification is good for the species, it is not a benefit that (under current theories) can comfortably be attributed to evolution. As Pinker points out, evolution is shortsighted and individualistic. Genes do not celebrate change for change's sake in hope of a payoff far down the road. In the short run, if a handful of mutant *Homo parametrus* were born into a group of *Homo rigidus,* they would have no advantage—not even a small one—within their own generation. On the contrary, they would listen to their *rigidus* relatives and learn the *rigidus* language, setting their parameters accordingly. Their actual linguistic behavior would be identical to that of *Homo rigidus,* and parameters would offer no selective advantage. Differences could appear only once a "parameter gene" became fixed in a population and that population was separated from *rigidus* long enough for its language to drift into another parameter setting. But any advantage that accrues to *Homo parametrus* in these circumstances cannot explain how the parameter gene gets established in the pool in the first place.

Individual genes might be able to derive immediate benefits from linguistic diversity if it acted as a reliable marker of kinship and biological relatedness. Within the "selfish gene" theory championed by Richard Dawkins and others, genes for "altruistic" behavior can do very well evolutionarily if the behavior is targeted toward the benefit of organisms that are likely to bear the same gene. But this works only if the "altruistic" organism can recognize those organisms that it is biologically related to. These considerations have led to various suggestions that the capacity for language differences evolved so as to promote "group solidarity" and to help us find good comrades in the struggle for survival. Language differences can certainly work this way in practice: Speakers of language X marry other speakers of X and kill speakers of language Y in many parts of the world. Semifictional language differences are even intentionally created for this purpose. For example, more weight is put on the difference between Hindi and Urdu since the partition of India and Pakistan, and it is no longer politically correct to speak of the Serbo-Croatian language, as it was only a few years ago.

Even if this is in some sense a good thing, it is not clear that it forms an evolutionary basis for the existence of parameters. The details still do not fit as well as one would expect. First of all, parameters are obviously overengineered for the purpose of group identification and solidarity. Simple differences in the pronunciation of words easily achieve this effect without involving the radical noninterintelligibility created by the parameters. For example, the biblical book of Judges records that the Gileadites identified their distant kinsmen the Ephraimites and marked them for slaughter based on whether they said *shibboleth* or *sibboleth*. The *s/sh* contrast is a trivial difference linguistically, but it was quite sufficient for their purposes. It is the same today with Serbian and Croatian or Hebrew and Arabic or "standard" English and African American Vernacular English. These variations mark ethnic barriers even though they do not differ in any major parameter.

In addition to this problem of overkill, an annoying detail regarding how children learn language is just as problematic for kin identification theories. In a relatively monolingual environment, one can maintain the comforting myth that children learn language from their parents. But in fact they don't: Children prefer to learn language from slightly older neighborhood kids. They also learn to communicate with their parents, of course, but their selection of primary language and even accent is guided more by their peers. This is why immigrant communities usually have a difficult time fully maintaining their language for more than a generation or two. It is also why language maintenance and revitalization programs run by Native American groups face such an uphill struggle. I remember the day I noticed, with some mild shock, that my daughter was speaking English with a distinct Montreal accent. This strong tendency in children is exactly the opposite of what one would expect if linguistic differences evolved to facilitate kinship identification. One would expect young children to focus on the exact speech style of their parents and other relatives and to avoid the style of the greater community—the opposite of what we observe. If you are worried about who your kids are, put down this book now and take a good look at their faces, because their voices may deceive you.

Steven Pinker is one of the few to frame these questions about linguistic diversity explicitly and to reject the obvious dead ends. "If the basic plan of language is innate and fixed across the species, why not the whole banana? Why the head first parameter . . . ?" he asks in the name of a hypothetical Martian scientist studying Earthspeak.

But when it comes to making a positive proposal with respect to this issue, Pinker and Paul Bloom acknowledge that "we can only offer the most tentative of speculations." Their principal suggestion is that some facets of language might have been so easy to learn with the cognitive apparatus that was already in place when a gene-based language recipe appeared on the scene that there was no pressure to specify those facets. For example, it is plausible to think that a very simple and general cognitive mechanism could recognize a fixed order between two designated elements, one a head and the other a phrase. If so, then leaving a head directionality parameter unspecified entails very little risk that the bearer of the gene will fail to learn the language. On the contrary, the correct order could easily be learned by observing a single sentence.

But Pinker and Bloom's suggestion has its own problems. First, in terms of my recipe analogy, it is odd to think that a parameter within a recipe could exist prior to the recipe itself. Perhaps my analogy makes this sound more paradoxical than it really is, but we must carefully articulate what we mean. Is a recipe that leaves open the direction of head and phrase really any easier to encode in a genome than one that specifies that order? It is not clear that it is. Specification of order might come automatically unless it is purposefully resisted, just as if I type two letters one will automatically come before the other, even when I do not attach any purposeful significance to the order.

Second, recent work has made it clear that the parametric theory does not make the problem of language acquisition nearly as trivial as one might think. Janet Fodor, for example, has emphasized the problems that come about because a child has to figure out many things all at once. It may seem trivial for a child who hears a sequence like "cookie eat" to infer that the language being heard is a head-last, Japanese-type language. But this is not at all a trivial matter if the child is not yet sure whether the sounds *cookie* and *eat* indicate nouns or verbs, whether *cookie* is the subject of *eat* or its object, or whether the sentence involves reordering to express focus (as in *My cookie, he ate*). All these matters and many others are still up for grabs in the first stages of language acquisition.

Third, the head directionality parameter is virtually the only syntactic parameter for which it is at all plausible to say that some general cognitive learning device apart from language could do the job. Many other parameters require a more sophisticated, language-specific deduction to set than this one does.

Finally, the Pinker-Bloom suggestion clearly takes parameters to be a kind of accident of evolutionary history. I disagree. I believe it is a design feature of human languages to have parameters strategically placed so as to maximize the apparent differences in E-language without much affecting the basic I-language. If this impression is accurate, then this aspect of language should be integrated into Pinker and Bloom's adaptive analysis, not just tossed in as a side effect.

In his more popular writings, Pinker offers a second hypothesis: Parametrization might have arisen so that learning can keep the speech of one person "in tune" with the speech of other people. The degree of learning required by having parameters could then correct for differences that would otherwise come into language because of random genetic variation (if indeed there are many genes available for this sort of thing, as he presupposes).

This suggestion should be testable linguistically. People should have detectable differences in their linguistic principles, but those differences should be compensated for by how they set their parameters. This means that two people who seem to speak the same language could actually have quite different internal recipes; they would be getting approximately the same linguistic effects by significantly different means. If that were so, then carefully controlled complex sentences should be able to reveal the difference. This issue has not been explicitly investigated in quite this way, so far as I know. But Ockham's razor cuts against it, and it runs counter to the deeply held working assumptions of most linguists. There is more individual variation in how speakers of a language respond to example sentences than we linguists sometimes like to admit. The differences, however, typically look more like differences in people's willingness to tolerate slightly suboptimal configurations, not true disagreements over which configurations are optimal. In general, the more one probes below the surface of people's grammatical intuitions, the more uniformity one finds, not the other way around, as Pinker predicts. And certainly there is no evidence (except in well-defined clinical syndromes) that the variation in peoples' linguistic responses is genetically based and subject to the laws of inheritance.

There are other ways that parameters might have a biological explanation even in the absence of a good adaptationist theory. Parameters might have been a biological accident rather than an adaptation. It could be that we have parameters not for the same reason that we have eyes, but for the same reason birds have colored plumage: Something that originally evolved to serve one function has been put to a new use. This would be the case if the mental structures (call them X) from which the general recipe for language evolved happened to have some property (call it P) that was the seed of the parameters. Thus, as X turned into the generic language recipe, property P automatically turned into parameters. Before we can fill out this picture, we must face another big puzzle in the evolution of language: What is the X that the language recipe developed out of? It is not easy to give a plausible answer. Perhaps the best suggestion to date is Bickerton's view that X was some kind of conceptual system—perhaps the one used by apes in perception, social cognition, or problem solving. There are some formidable difficulties for such a view, and the more one knows about the details of grammar, I think, the more formidable they seem. We can put these concerns aside for now, however, because the crucial question for us is not what X is but what P is. What property could the primate conceptual system have had, such that when that conceptual system developed into a language system, the result was parameters? It is hard to see. What reason do we have for believing that *conceptual systems* contain parameters of some kind? Is the language of thought parametrized? That seems unlikely. Even if it were true, it would only push the interesting question back one step. Why should conceptual systems allow for this kind of variation?

Finally, parameters might exist because of physical necessity. It could be that some law of physics happens to have the consequence that a language recipe cannot be specified in the brain without having parameters. Since we have so much ignorance about these matters, it is impossible to say for certain that it could not be so. But the claim does not sound very plausible given the current state of our knowledge.

Overall, then, parameters seem to be a salient property of human language, but one whose existence is not readily explained by evolutionary biology as we know it. Some people might take this as a refutation of the very idea of parameters, on the grounds that everything must be explained by evolutionary biology. This seems to me a dangerous argument. It would deny the reality of what we can study rather directly in favor of what we cannot observe directly (such as earlier stages of language in extinct hominids). Certainly there is nothing internal to evolutionary theory that guarantees that it is the explanation for everything. Nor do those who deny the reality of parameters on evolutionary grounds have a viable alternative account for the kinds of facts we have been considering throughout this book. For the time being, it seems better to live with the puzzle than to deny it.

———————

They say that before you marry someone, it is wise to meet their family and other close relatives. Partly you need to know those people for their own sakes, because they will become your family, too. But a second reason is that you get to know your prospective spouse in a new and valuable way, by getting a taste of the environment he or she grew up in. Of course, it is a mistake to infer that your intended is an exact duplicate of the people you meet on such a visit. Nevertheless, if you are alert, you are almost certain to get some new understanding of who he or she really is.

We can try this same methodology to seek some positive new understanding of what parameters are and why they exist. Rather than hastily shoehorning them into a familiar explanatory niche, we can take a closer look at how they interact with other factors to produce different languages in practice. Even though we were not there when parameters were born and did not see them in early childhood, we might get some insight into why they are the way they are by considering who their relatives are and how they relate to them. With this in mind, I take a brief look at the dynamics of how languages change from one type to another in the course of their history, seeking some hints about the nature of parameters.

However we relate to the Genesis story of Babel, we know that direct divine intervention is not needed to produce different languages now. Languages have changed and diversified within recorded history, and we have documents that enable us to trace some of those changes. For example, Classical Latin developed in different ways in different parts of the Roman Empire to give us French, Spanish, Italian, Romanian, and other modern languages. Similarly, Sanskrit turned into Hindi and many of the other languages of the Indian subcontinent. Even in parts of the world where written records of language do not go back far, we can infer some of their historical development by studying the patterns of similarity and difference among related languages. Some of these changes (although not all) involve changes in the settings of the parameters we have been discussing in this book. I have already mentioned that the languages of north central Australia are polysynthetic and other languages of Australia are not, even though they all show signs of descent from the same ancestral language. Thus, one or the other of the stocks must have changed its setting for the polysynthesis parameter. Similarly, French changed from a null subject language (like Latin) to a non–null subject language and lost its case marking. Hindi became a head-final language, unlike most other Indo-European languages, and has developed a partially ergative case-marking system. Old English was like Modern German and Dutch in that it was a mixed subject-object-verb language with extensive case marking. But by the Middle English period, English lost its case marking and changed its word order to the point that now it is a prime example of a head-first, subject-verb-object language. In short, the settings for parameters can change.

The process by which languages change is reasonably well known and is described in many books. This process has often been compared to the way new species evolve—initially by Darwin himself—although there are important disanalogies as well. Here I sketch the process only in broad terms, so as to highlight a few aspects that seem particularly relevant.

The first step in the development of a new language is for a group of speakers of the protolanguage to become isolated for geographical or social reasons. For example, the transportation system that linked the disparate regions of the Roman Empire broke down, the Angles and the Saxons invaded the island of Britain, and the landowners of Caribbean sugar plantations refused to speak to their new slaves. Next, innovations come into the language of the isolated group that are different from the innovations of the main group. For example, a new way of hauling goods is invented, and the people on one island call them *trucks* (because they have wheels), whereas the people on another island decide to call them *lorries* (because they pull things). These changes and additions are passed to subsequent generations by the normal course of children learning language from their parents and older children. For example, parents on both islands might equally take their toddlers into their laps and play "what's that?" games with board books, but for some families the right answer for the thing on page 10 will be *lorry* and for others it will be *truck*. In this way the innovations of each generation can accumulate, so that eventually the speeches of the two communities have a markedly different flavor. Finally, these innovations can reach a critical mass in one group or the other, so that most children unconsciously decide that the patterns are best formed by using a different parameter setting than their predecessors used. In effect, they reanalyze the old exceptions as the new rule and the old rule as the new exceptions. They then tend to regularize the language around a different core of "well-behaved" cases, and pass their view on to younger people. At this point the language not only has some distinctive vocabulary and turns of speech but is actually formed by a different recipe than the original language.

What clues about parameters can be identified in this process? There is little of linguistic interest to be gleaned from the brute fact that groups of speakers become separated by migration or social distinctions. But the appearance of innovations in the language after separation is very significant. It is crucial to note that these innovations are not entirely predictable from the initial properties of the language being spoken. If they were deterministic, then the same innovations would come into both speech communities, and their languages would not diverge. Rather, these innovations play for linguistics approximately the same pot-stirring role that random mutations do for biology. What, then, are these innovations, and where do they come from? The example of *truck* and *lorry* was a simple case of coining different words for some item, but this is only a small part of the answer, since innovations affect more than just vocabulary.

Innovations also come into language as a result of speakers' style choices. I have often spoken of language recipes as rules that must be followed or constraints that must be obeyed if one wants to construct a sentence of (say) English. This talk of rules and constraints might sound stifling to creativity, but in fact it is quite the opposite: They are what make an important kind of creativity possible. Rules give freedom. Just as trains that stay within the limiting confines of railroad tracks can get to any city in the country quickly, so speakers who stay within the limiting confines of grammar can effectively communicate whatever they want. It is precisely because English has rigid subject-verb-object word order that newspapers can write "Man Bites Dog." Mohawk does not have fixed word order, but it communicates the same difference through rigid rules about having agreement markers for subject and object on the verb. This underscores the point I made earlier, that the ability to speak different languages does not increase biological fitness by increasing the number of things one can talk about. Each language contains the flexibility to talk about virtually anything. More than that, all languages make alternative versions of sentences available, which speakers use to achieve stylistic effects. For example, in addition to a stylistically neutral sentence like *The man bit the dog*, one can say any of the following in English:

The man bit the *dog*.	(focal stress)
That dog, the man bit it.	(dislocation)
The dog was bitten by the man.	(passive voice)

All three of these communicate the same basic fact as the original, but one can use them stylistically to emphasize the dog, either by saying it louder or by saying it first. We might do this if the dog is the crucial piece of new information, or if the dog is the topic of our discussion, or if we thought it would make us sound interesting and dynamic. One significant kind of innovation, then, could be a new fashion governing which version of a sentence one is inclined to use for a particular effect.

The importance of these stylistic choices becomes clearer if we look at a case study of how English changed from the subject-object-verb language of *Beowulf* to the subject-verb-object language of Chaucer and Shakespeare. At first glance it seems as if such a change should be impossible. How could a language shift from one word order to the opposite word order without going through a stage in which the children of one generation truly cannot understand their parents (as opposed to merely pretending they don't hear)? But the change did happen, and it is well documented; David Lightfoot in particular has studied it from a parametric perspective. The crucial point is that changes in the use of a stylistic rule prepared the way for the dramatic change in parameter.

Simplifying a bit, Old English typically had Japanese-style word orders:

The man the dog bit.

It developed a stylistic rule, however, in which the verb could be fronted to a position before the subject. This happened particularly when the sentence was introduced by a conjunction, resulting in forms like:

. . . and bit the man the dog.

Such sentences became more and more common in simple main clauses—which are the kind that children pay attention to. Now put yourself in the position of a child trying to learn Old English. Suppose you can recognize (somehow) that this must be a stylistically derived order, with the verb moved out of its normal position. (Perhaps such sentences have a special intonation, or you realize that they violate the verb-object constraint.) Even so, you do not know where the verb was before it was fronted. It could have started out at the end of the sentence (as the adults in fact believe) or in the middle of the sentence. In other words, *and bit the man the dog* could have been a stylistic variant of:

The man bit the dog.

This, of course, is the Modern English word order. The stylistic rule thus creates some indeterminacy in what the child can infer about basic word order from common, everyday sentences. To further muddy the waters, Old English adults could put a special topic phrase before the fronted verb, as still happens in the other Germanic languages. This topic phrase could be an adverb, for example:

Yesterday bit the man the dog.

But a very common choice for the topic of a sentence is the subject. Thus, the last Old English speakers would say things like:

The man bit the dog.

For them, this sentence has the verb fronted from the end and the subject fronted from the middle of the sentence. But their children, who were eagerly looking for evidence by which they could set the head directionality parameter, naturally seized on such sentences. Mistaking them for underived sentences, they said "Aha! English is a head-initial language." And they became self-fulfilling prophets, since they spoke English so and taught their descendants to do the same. When the fad for fronting verbs eventually died out, English would be left with the unambiguous subject-verb-object word order we know and love. The essential point is that the transition between the two parameter settings, which otherwise would have been impossibly abrupt, was smoothed by the "stylistic" verb-fronting rule. It was the frequency of this rule that allowed the last Old English parents and the first Middle English children to understand each other.

Similar changes in the frequency of using optional, "stylistic" rules underlie many of the other parametric shifts that we know of. Consider, for example, what might happen if the speakers of a language became enamored of the passive voice. Intransitive sentences like *The dog barked* would look the same as in other languages. Children would rarely hear *The dog bit the man,* however, usually hearing instead *The man was bitten by the dog.* These children would notice that the semantic object of most two-character sentences (here *the man*) is expressed in the same way as the subject of the intransitive verb, because (from the adult perspective) it was made into the subject by passive. The children would also notice that the understood subject of two-character sentences (here *the dog*) is flagged by a special marker, *by.* These are the defining characteristics of an ergative language, such as we saw in Chapter 6. This illusion would be particularly compelling in a head-final language, where the verb comes at the end in both active and passive sentences and where a case-marking system is expected. In this way the frequent use of the passive led many Polynesian languages to become ergative languages. The opposite change seems to be happening in arctic Quebec varieties of Eskimo, where frequent use of antipassive (the inverse of passive) is shifting what has been an ergative language into an accusative one. Similarly, dislocation sentences like *That dog, the man bit it* in languages like English are relatively rare; they are used only for particular effects in certain speech styles. But equivalent dislocated sentences have become the norm in many colloquial varieties of the Romance languages. This tendency, together with the object pronoun sticking onto the verb (as they tend to do) could cause children to analyze the Romance languages as they do Bantu languages like Swahili, with the optional polysynthesis parameter set positively. Additional changes of this type could even lead to a true polysynthetic language.

The question, then, boils down to why speakers change the way they use these optional, stylistic rules, which are the engines of grammatical change. The answer is that they just do. Rigorous principles determine aspects of the form of what people say, but not the content. At this point we cannot avoid the importance of people's free will as expressed by their language use. Our language recipes are so powerful that they make available an unlimited number of possible sentences that we could utter on any given occasion. And we take advantage of this freedom, rarely saying the same sentence twice except in certain simple and formulaic situations in which we predictably say, "How are you?" or "May I help you?" or "_____ you!" (expletive deleted). It is generally impossible to foresee in any detail what sentence people will speak on the basis of any reasonable description of their history or current situation. Certain broad statistical tendencies emerge, of course, but these can be violated at will. Furthermore, although the use we make of our language recipes on a given occasion does not seem to be determined, neither is it merely random. Rather (with some notable exceptions), we judge what people say to be *appropriate* to the occasion. People's speech—including their use of stylistic options—is purposeful, accomplishing rational goals in the context of the situation. Chomsky has often emphasized that human speech is unbounded, stimulus-free, and appropriate. In this he follows philosophical arguments going back to Descartes, who claimed that any behavior that had these three properties could not be explained in mechanical terms. (To entities with this sort of behavior, Descartes argued that it was reasonable to attribute an immaterial soul.) Be that as it may, there is no denying that we have at least the illusion of a free will, and this free will expresses itself clearly in the way we talk. Apparently, even higher authorities than the U.S. Constitution guarantee us a kind of freedom of speech. The exercise of this freedom produces the innovations and statistical variations that drive language change, including changes in parameter settings. Free will, then, is the first relative of the parameter that we can identify.

The second crucial factor in language change and diversification is that children learn particular language recipes from the people around them. If it were not so, then every child would invent a new language from scratch, and widespread communication would be impossible. That language is learned is a trivial observation. The nontrivial part is that children learn language *from people*. They may hear language from many other sources: televisions, radios, CD players, computers, or talking Elmo dolls. Some children get many hours of this kind of linguistic input. Yet it is striking that they do not learn language from this input—or at least not much. Studies have been done of Dutch children who heard many hours of German TV to assess whether they learned German from this exposure. The answer is no. Deaf parents used to be encouraged to let their hearing children watch lots of television so that they would learn English, but this teaching strategy proved as ineffectual as it was easy. My own children spent their early years in Montreal, where they heard French spoken all around them. Nevertheless, they managed to avoid learning even a word of French until they went to school and actual people started talking *to them* in French. Why is this so? Why does television, which we consider such an effective teacher of violence and materialism, fail us in the simple task of teaching language?

There may be many facets to this question, but I believe that a crucial consideration centers on the intentionality of language. Our sentences not only have grammatical properties, such as heads coming before phrases and verbs agreeing with nouns; they also have semantic properties. They mean something. They represent propositions about the world that can be true or false. As a conscious mental exercise, it is possible to construct sentences that obey the rules of English but do not mean anything, as Lewis Carroll did in his poem "Jabberwocky" or as in Chomsky's famous example *Colorless green ideas sleep furiously*. But such sentences are clearly not the norm. Our attachment to semantic content in language goes very deep. It is impossible to listen to someone speaking a language that you know and notice the grammatical structure of their sentences but not attend to their meaning. Intentionality is probably also crucial to language acquisition. Children will pay serious attention to a language and make progress in cracking its code only if they have some access to what the sentences that they hear mean, in the form of interactions between them, the speaker, and their shared environment. This is where a language-rich but interaction-poor medium like television fails: It does not provide a young child with enough salient clues as to what the blips on the screen are talking about or why we should care.

One could imagine its being otherwise. One can imagine someone's learning a language by doing some kind of sophisticated statistical analysis to the point that she could reliably produce novel sentences in the language even though she had no idea what those sentences mean. But it is clear that this never happens in practice, and it is hard to imagine how it could. We simply cannot learn a language without knowing what it means. This is true even for intellectually sophisticated adults such as the Japanese cryptographers who tried to decipher Navajo. Similarly, classics scholars were unable to crack the code of Egyptian hieroglyphics, despite having access to abundant inscriptions, until the Rosetta stone gave them some idea of what one of those inscriptions was talking about. The intentional, meaning-bearing nature of human language is thus a second relative of the parameter.

A third important factor in the process of language change is the way children jump to conclusions when learning a language. This normally allows them to learn perfectly a language they are interacting with, in spite of frequent gaps in the samples they are exposed to. In some special cases, however, they may end up learning a significantly different language recipe from the one that is actually producing the sentences they hear. The example of word order change in English discussed above is a good illustration of this, too. Children know that the verb must start out next to the direct object, even though they often hear it at the front of the sentence because of stylistic verb fronting. They know this prior to—even in spite of—exposure to the language because their innate recipe for language includes the verb-object constraint. They also know that the verb should either come consistently before the object or consistently after the object because they know that the recipe for language must set the head directionality parameter one way or the other. Therefore, when they hear a critical mass of sentences like *The man bit the dog,* with the verb coming immediately before the object, they jump to the conclusion that the language they are learning is a head-first language. From the perspective of the actual adult grammar, this may not be true, but the difference is not apparent in the simple sentences that the child attends to. Thus, the language changes, and remarkably quickly. Many readers of historical texts have observed that once the frequency of verb-object order in main clauses reached a certain point, the basic word order of English changed radically within a generation or two, around 1100. (One can observe this abrupt change by looking at embedded clauses, where the obfuscating process of verb fronting did not apply. Before 1100, sentences like *Hrothgar said that the man the dog bit* were common, and after 1100 they were replaced by sentences like *William said that the man bit the dog*. Transitional sentences like *Hrothgar said that bit the man the dog* [with verb fronting] were never common. Although the word order in main clauses changed gradually, the word order in embedded clauses changed suddenly at the point where children started jumping to the opposite conclusion.)

In the same way, when Polynesian children heard a critical mass of passive sentences in which the agent of a two-player verb was marked differently from the agent of an intransitive verb, they jumped to the conclusion that the language was ergative. This caused it to be ergative from then on.

If the past few decades of research in cognitive science have taught us anything, it is that this tendency to jump to conclusions in ways that are determined by innate knowledge is pervasive. We see an ellipse widen into a circle, flatten into a line, then widen again. This pattern could have been made by any number of things, but we perceive it as a rotating disk. We see some patches of color on a two-dimensional page, and we say it is a lion. We come to know about perfect circles and parallel lines and the number five, even though we have never sensed even one of these things. We reason on the basis of a natural logic without being explicitly trained in it. In all these domains, we combine a hopelessly degraded and underspecified stimulus with some innate knowledge to reach our conclusions. And what's wonderful is that we are almost always right. We become aware of how much we are adding to the basic stimulus only in special circumstances that are manipulated in the psychology labs and science museums to create optical illusions. The way we learn language fits this pattern exactly: We combine the limited and imperfect sample of sentences that we hear with our innate knowledge of the recipe for human language to deduce the recipe for the particular language spoken around us. And usually we are right—except in those special cases in which sentences with ambiguous structure lead to a parametric change. Our capacity for innate, a priori knowledge is then the third essential relative of the parameter.

We have identified three relatives of the parameter that crucially participate with it in creating a diversity of languages: free will, intentionality, and a priori knowledge. The three have something in common: They are all on the list of the great mysteries of philosophy and cognitive science. I have a special sense of the word *mystery* in mind here—one introduced by Noam Chomsky, extended and amplified by the philosopher Colin McGinn, and endorsed by the psychologist Steven Pinker. These authors distinguish a mystery from a puzzle. Both are questions that have not been answered by a scientific theory. The difference is that a puzzle looks like it can be solved by scientific research progressing along its normal channels. A puzzle has not been solved *yet,* but it resembles other questions that have been solved. One can see ways of understanding aspects of the phenomenon in question that can potentially be pieced together for a complete theory. Sometimes a puzzle remains because there are two or more good theories that address it, and it is unclear which is best.

With a mystery, one cannot imagine how to get started using the tools of scientific inquiry. It seems to have a qualitatively different character from questions that have been explained in the past. There is no sense of making progress on a mystery, even though it may have been a focus of thought for millennia—as some of these topics have been. There seems to be no way to break a mystery down into simpler, more manageable pieces. Mysteries are left open not because they are explained by two or more rival theories but because it is hard to imagine how anything could be a theory.

Because of their resistance to scientific explanation, McGinn writes, mysteries are often given quasi-mystical or religious significance, including an appeal to some new, nonphysical force. Descartes's invocation of an immaterial soul to account for free will in language use would fall into this category, as would many modern proposals to make consciousness a primitive feature of the universe. Alternatively, the mysteries are often denied altogether. Tough-minded people argue that free will, intentionality, and a priori knowledge are illusions or confusions, the only reality being neurological events in the brain. In that case there is really nothing for science to explain. (Of course, one would still like an explanation of why these illusions arise, and why they are so universal and persistent that it is impossible for us to live without them. . . .)

As an alternative to both mysticism and skepticism, Chomsky, McGinn, and Pinker suggest that mysteries have perfectly ordinary scientific explanations in principle, just as puzzles do. The difference is that those explanations happen to be outside the range of what the human mind can grasp. On this view, reaching the correct theory of free will and intentionality is impossible for us for the same reason that prime factorization is impossible for dogs, however intelligent they might be at things like herding sheep.

That these mysteries have never been solved is of course no guarantee that they never will be. They may turn out to be puzzles after all. Perhaps because I have been part of an intellectual enterprise that has made measurable new progress on age-old questions, I am reluctant to infer future failure from past failure, even while honoring the greatness of those who have failed. After all, the two great unsolved problems of my mathematical youth, the four-color map theorem and Fermat's last theorem, have both been proved in recent years. Anything is possible.

I find it intriguing that all the mysteries that McGinn enumerates have something in common besides their history of defying analysis.

They all concern fundamental questions about how the human mind relates to the external world. The mystery of free will is how a mind can affect the world in a way that is neither random nor determined by physical causes. The mystery of intentionality is how thoughts in the mind can be said to represent truths about the world, even though they do not resemble those truths in any obvious way. To think that a state of affairs is true, for example, is not *literally* to have a picture of it in one's head. No such pictures have ever been seen in a brain autopsy. Finally, the mystery of a priori knowledge is how the mind can know things about the world—especially abstract universal truths— of which it has had no adequate experience. For example, we know with certainty that the square of the hypotenuse of a right triangle is equal to the sum of the squares of the remaining sides. This is an extremely practical and well-tested piece of knowledge, which much engineering is based on. Yet how is it that we can know it when none of us has ever sensed even one perfect right triangle?

The three mysteries are also interrelated in such a way that progress on any of them might constitute progress on the others as well. For example, we would not be able to act freely if we were unable to think thoughts about the world around us. How could one act purposefully and appropriately in the world if one had no idea what the world was like? Furthermore, our a priori knowledge of the world is a special case of our meaningful ideas of the world, providing the necessary grounding for the other, more contingent ideas triggered by experience. Thus, all three mysteries fall in the same general domain. This is not what one would expect if these questions were simply the leftover parts of the world that happen to be beyond the reach of human intelligence. One might expect human intelligence to reach its limits on a disparate variety of topics that have nothing intrinsically to do with each other—just as my children might make mistakes on a variety of widely different topics in their year-end school tests. Of course, the similarity and interrelatedness of the mysteries could be a consequence of the fact that the people involved in these discussions are all cognitive scientists and thus tend to think about similar topics. Yet perhaps we are missing just one fundamental idea—something unforeseen about the relationship of the mind to the world—that could be the key to all these mysteries.

In any case our understanding of the nature of human beings seems to be short at least one major idea, either because the idea is beyond our collective ken or because no one happens to have had it yet—or because we paid no attention to the one who did have it. Since we have a known gap in our knowledge, it is not so surprising that we find no clear reason why human languages should have parameters. There is no need to force our understanding of parameters into either a cultural or evolutionary slot until we know what other categories for explanation are available. This does not necessarily mean that the existence of parameters is itself one of the great mysteries of cognitive science. Perhaps it is a specific subcase of the bigger mystery of a priori knowledge. Or perhaps the innate language recipe is not intrinsically mysterious but has been pushed toward having parameters by other factors that are mysterious. If so, then understanding the whys of parameters would be like understanding why a pond in the mountains has a peculiar irregular shape. Understanding rock shapes is extremely relevant to understanding pond shapes, not because ponds are a type of rock but because rocks push ponds into their shapes. Understanding the mysteries of cognition could be a necessary precondition to understanding parameters in more or less the same way.

———————

What value, then, should we place on the differences among languages? This is a matter of considerable emotional and practical importance. On the one hand, local linguistic diversity has increased in many places, thanks to migration and improved transportation, raising questions of how we can all get along. On the other hand, the linguistic diversity in the world as a whole is decreasing very rapidly. Approximately half the languages currently spoken are on the verge of extinction, and many more are at risk. This is a pity from my point of view, given that I make my living studying different languages. It is also sad from the point of view of many of my friends and colleagues, who as speakers of these languages feel a deep sense of loss at their seemingly imminent demise. But how should society as a whole look at this issue? Is the loss of linguistic diversity something to be mourned and prevented if possible by research, language planning, and education programs? Is it something to rejoice in, perhaps even to help on its way by assimilatory policies and programs? Or is it a matter of no great concern? The answers to these questions depend in part on factual matters about how languages differ.

Suppose one thought that the differences between languages were not very substantial, that languages vary only in the words that they have for things, plus some minor differences in how the words are arranged. If languages are not fundamentally different from one another, then obviously they do not differ in value. One language is as good as another for any purpose one can imagine. People's attachments to the languages that they speak and their use of language in identifying who they are would be no different from their attachments to a particular national flag. These feelings might be powerful, but they would not be deep or principled. One flag is not inherently any better than another, except perhaps aesthetically. Linguistic pluralism on this view would be nothing but a surface veneer, and an inconvenient and expensive one at that. In Canada, which is officially bilingual between French and English, one must find room for both languages on every cereal box, one must translate every document, one must learn another language for any service-oriented job, and one is uncertain even how to greet someone in the street. These tactical problems are compounded if two languages become three or more. If language differences have no value beyond an emotional one, and if they are expensive and inconvenient to maintain, then it is easy to conclude that linguistic diversity is not worthwhile. Let them all learn English.

Suppose, in contrast, languages are fundamentally different, that one language can express ideas that cannot be adequately translated into another. Then the question of value does arise. Specifically, one could think of the value of one language compared to another as being proportional to the value of those ideas that can be expressed in the first but not in the second. We are used to the notion that ideas have value, even if it is not always clear how to quantify it. This is essentially what Nowak did in his mathematical study, where value is cached out evolutionarily as the probability of reproducing. This is also the view that most supporters of linguistic diversity (and pluralism more generally) seem to have, at least implicitly, although they are vague about just what the value is. Although this view can provide a basis for saying that minority languages should be supported and promoted, it also can awaken certain fears. Racism and prejudice can be stimulated by any claim that the cultural equipment of one group is superior to that of another group in some respect. There is plenty of prejudice already, without linguistic support. The idea that one language might be considered better than another in some global sense will therefore be repugnant to many readers—and rightly so. It is not easy to rule out the possibility in an intellectually sound way, however, given the premise that languages differ in their expressive powers. Usually people avoid racist conclusions by assuming that one language is better in one domain and another language is better in a different domain, so that the overall value of one is not greater than that of another. But it would be very difficult to prove this. On the contrary, it would be as surprising an outcome as creating two columns of random numbers and discovering that both columns sum to exactly the same total. I believe that we need a sounder basis for our valid intuitions that all languages are of equal value.

Moreover, if different languages have value because they can express ideas that cannot adequately be translated into another language, speakers of one language will inevitably and systematically fail to understand speakers of another language in deep ways. This could lead to an inability to achieve meaningful agreement, various kinds of social breakdown, mistrust, discrimination, and even one group's imposing itself on another with naked force. This is the dark side of pluralism. The danger of fundamental misunderstanding should be directly proportional to the degree to which languages are different, and hence to their potential value. Even if everyone practices self-restraint, tolerance, and courtesy, there would be little possibility of one linguistic group's enriching another simply because there is little possibility of real understanding between them. On this view, the best we could hope for from pluralism is peacefully to live our separate realities.

The parametric theory of languages, however, insists that neither of these simplistic views of linguistic diversity is exactly correct. Languages are significantly different but commensurable. They vary widely in their visible sentences but are very similar in their recipes. Mohawk sentence structures are unlike those of Japanese, which are in turn unlike English, but the differences are systematic and predictable. Thus, a Japanese sentence can be matched to an English one by an algorithmic process of reversing the order of heads and phrases. An English sentence can be matched to a Mohawk one by systematically dislocating the noun phrases. The problems increase when several parameters interact at several levels and when lexical differences are thrown in as well. Still, translation from one language into another should be difficult but possible. This is what the Code Talkers showed us.

This view of linguistic diversity points the way toward a more solid positive attitude toward linguistic pluralism. It means we can maintain a diversity of languages without engendering serious misunderstanding—so long as we keep making the effort to understand each other. We can also give value to different languages without inviting the claim that one language has more value than another. Even though all languages can express all ideas, this does not mean that all languages express those ideas with equal ease. It is easy to find instances in which what is said simply and directly in one language requires a lengthy circumlocution in another language. Some languages have verbs that make detailed distinctions of tense, others have verbs that automatically mark the direction of motion, still others have verbs that mark the posture of the participants in the event. Some linguists have conjectured that the need to express one or the other of these notions grammatically in a language may cause speakers to pay more careful attention to those features. This in turn could cause them to remember those aspects more clearly. These conjectures raise the intriguing possibility that some ideas worth having might occur more easily to people of one linguistic group than to another.

Second, even though all languages may express the same propositions, they clearly will not all support the same poetries. Verbal art of all kinds, from the heroic epic to the playful pun, depends on idiosyncratic correspondences between sound and meaning. For example, the elaborate interlocking rhyme scheme of Dante's *Divine Comedy* is greatly facilitated by the tense and agreement suffixes that go on Italian verbs to make Italian a null subject language. Because many words end in the same suffixes, Dante had a particularly large stock of rhyming words to work into his poetic scheme. The iambic pentameter of Shakespeare and Milton's poetry, for its part, is well suited to English, which has heavily accented syllables that determine the timing of speech. Other kinds of languages invite a different poetry: the subtle nuances of word order in a Mohawk oration or the effects of perspective achieved by logophoric pronouns and ideophones in a West African folktale. The world is a richer and more beautiful place because of its diversity of languages.

From this perspective, the benefits of linguistic pluralism can be compared to the benefits of having a plurality of eyes. Look at the things around you in the middle distance, and close first your right eye, then your left eye. The view from your right eye is not inherently better or worse than the view from your left eye. Indeed, the view from the two eyes is not very different in objective terms. Nevertheless, there is great value in having two eyes, and we do not think that one of them is redundant and dispensable. The reason is that when your neural circuitry compares the two views, sight takes on a new quality: The world becomes three-dimensional. As a society we can expect to use linguistic diversity in the same way.

Notes

CHAPTER 1

1   Navajo Code Talkers in World War II: Bernstein 1991; Paul 1973; Johnson 1977.

3   Early predictions about computers and language: Turing 1950; Newquist 1994.

7   Navajo words, sounds, and prefixes: Young and Morgan 1987.

9   Navajo subjects: Speas 1990. Navajo plowing: Young and Morgan 1987. Navajo dictionaries: Kenneth Hale's preface to Jelinek, Midgette, et al. 1996.

9   Navajo word order: Speas 1990; Young and Morgan 1987.

10   *Yi-* and *bi-*: Hale 1973; Speas 1990; Young and Morgan 1987.

11   Quickness and accuracy of code talkers: Bernstein 1991; Paul 1973.

13   Universal grammar as key to language learning: Chomsky 1975: ch. 1; 1980; and many of Chomsky's other writings in passing. See Pinker 1994 for a good general discussion.

15   Navajo similarities to English: Young and Morgan 1987; Speas 1990.

16   *Yi-/bi-* like the passive: Chomsky 1981.

18   Language, thought, and culture: There is an enormous literature on these matters from various disciplines. See, for example, Whorf 1956; Gumprez and Levinson 1996.

CHAPTER 2

21   Greek chemistry: Gaarder 1994.

23   Parameters for language differences, learnability: Chomsky 1981. The quotation is from pp. 3–4. Chomsky's idea of a parameter is only distantly related to his famous notion that languages have a "deep struc-

ture." The deep structure proposal is that the sentences of a language start out in a certain canonical form, and this gets modified by transformations. Thus, it concerns the abstract similarities among different sentences in a single language, whereas the parameter proposal concerns the abstract similarities among the grammars of different languages. The reasoning is somewhat similar, but the level of analysis is different. The logical independence of the two ideas is underlined by the fact that in his recent writings Chomsky abandons the notion of deep structure as such but maintains (a version of) parameters: Chomsky 1995.

24 Alchemists and early chemists: Brock 1992; Ley 1968.

25 Boas, Sapir, and Bloomfield: Campbell 2001; Robins 1989; Stocking 1974. Bloomfield's obituary of Boas: Bloomfield 1970.

26 Nootka and Paiute: Sapir 1921. On Nootka nouns and verbs, see also Sapir and Swadesh 1946.

27 Word orders in South America: Derbyshire and Pullum 1981. Percentage of languages in New Guinea: Grimes 1988.

28 Döbereiner's triads: Atkins 1995.

28 Greenberg's word order universals: Greenberg 1963. Navajo examples: Young and Morgan 1987. Japanese examples: Kuno 1973.

30 Edo word order: Agheyisi 1990; also my own joint research with O. T. Stewart.

31 Other early classifications of elements: Atkins 1995; Brock 1992. 625-language sample: Dryer 1992.

33 Head marking and dependent marking: Nichols 1986, 1992.

33 Japanese examples: Kuno 1973. Mohawk examples: Baker 1996; see also Mithun and Chafe 1979.

35 Dalton: Brock 1992; Ley 1968; Atkins 1995.

36 Italian vs. French and English and the origins of the parameter: Chomsky 1981; Rizzi 1982; Kayne 1984a.

39 The observation that languages with no need of pronouns also allow free questioning of subjects was originally due to Perlmutter 1971. The further correlation of these two properties with allowing verb-subject word order is due to Kayne 1984b.

39 Spanish like Italian: Perlmutter 1971; Jaeggli 1982. Romanian like Italian: Dobrovie-Sorin 1994.

44 Parameters and parametric clusters: Chomsky 1981; Rizzi 1982.

45 Reactions to Chomsky: see, for example, Givón 1979; Foley and Van Valin 1984.

47 Conditions prior to Mendeleyev: Atkins 1995; Ley 1968; Brock 1992.

49 Virtues of the periodic table: Atkins 1995.

CHAPTER 3

54 I-language and E-language: Chomsky 1986b.

59 Japanese word order vs. English: Kuno 1973 and many others.

62 Worldwide distribution of two language types: Greenberg 1963; Dryer 1992; Nichols 1992. Edo word order: my own research with O. T. Stewart. Lakhota word order: Williamson 1984.

62 Percentages of languages with each word order: Tomlin 1986.

66 Phrases and phrase structure trees are discussed in most linguistics texts. Radford 1981 is particularly good on these matters.

67 Piecemeal sentence construction: Chomsky 1995.

68 Head directionality parameter: This version is a synthesis and cleaning up of various versions found in the literature. Its conceptual origins are in Chomsky 1981 and Stowell 1981, as extended to functional categories by Chomsky 1986a. The overall presentation follows Dryer 1992 most closely.

72 Articles as heads in phrase structure: Abney 1987.

73 Alternative word orders are sometimes possible in both languages, to create a special effect. Thus, one can say *Three books I will buy* in English as well as the more normal *I will buy three books* when the focus is on the object. Similarly, object-subject-verb order is a rather common alternative in Japanese when focus is on the object.

75 Tzotzil word order: Aissen 1987. Malagasy word order: Guilfoyle, Hung, et al. 1992; Keenan 1976; but see Rackowski and Travis 2000 for a reanalysis. Hixkaryana word order: Derbyshire 1985.

78 Optimality theory in general: Prince and Smolensky 1993. Applied to syntax: Grimshaw 1997, among others.

79 Variable complementizer positions in Nupe: Zepter 2000.

80 Parameters as lexical knowledge: Borer 1984; Fukui and Speas 1986; Chomsky 1995.

80 Formalism vs. functionalism in linguistics: Newmeyer 1998; Croft 1995; and many others.

81 Explaining word order generalizations by parsing ease: Hawkins 1990, 1994. Many others, including Dryer 1992, make informal remarks in this direction.

81 Prepositions evolved from verbs: Lord 1993.

82 Amharic word order: Amberber 1997. Statistics on verb-final languages with prepositions: Dryer 1992.

84 Survey of null subject properties: Gilligan 1991. The matter is also discussed in Newmeyer 1998.

The Atoms of Language

CHAPTER 4

86 General characteristics of Mohawk: Mithun and Chafe 1979; Baker 1996. Oneidas in the army: Bernstein 1991: 46.

87 History of terms *synthesis* and *polysynthesis:* Foley 1991; Robins 1989; Whaley 1997.

88 Teaching Mohawk words: Deering and Delisle 1976.

89 Nonconfigurationality in general: Hale 1983; Speas 1990. Nonconfigurationality in Mohawk: Baker 1991, 1996.

91 Noun incorporation in Mohawk: Baker 1988, 1996; Mithun 1984.

92 English compounds: Selkirk 1982; Lieber 1983; and many others.

93 Interpretation of English compounds: Selkirk 1982; Roeper and Siegel 1978.

93 Restrictions on noun incorporation in Mohawk: Baker 1988, 1996; Mithun 1984.

93 This statement of the verb-object constraint is modeled most closely after Selkirk 1982 (who calls it the "first order projection condition"). Some version of such a statement is a common ingredient of syntactic theories; see also the uniformity of theta assignment hypothesis of Baker 1988, 1997.

96 Causatives in Mohawk: Baker 1996: ch. 8.

97 Parallels between causatives and noun incorporation: Baker 1988, 1996.

98 Nonconfigurationality in Mohawk: Baker 1991, 1996; Mithun 1987. Completeness condition: Chomsky 1981; Bresnan 1982. (The terminology comes from Bresnan; Chomsky refers to this idea as the "projection principle.")

99 Pronoun omission related to verb inflection: Rizzi 1982; Jaeggli and Safir 1989.

100 Mohawk verb inflection: Lounsbury 1953; Postal 1979.

102 Dislocation in English and Romance languages: Cinque 1990.

103 Free word order in Mohawk as dislocation: Baker 1991, 1996. In these works I built on a fairly similar proposal by Jelinek (1984) for other languages.

105 Reflexive sentences: Chomsky 1981; Reinhart and Reuland 1993.

106 Reflexive sentences in Mohawk: Baker 1996.

108 Dislocation of quantifiers: Rizzi 1986; Cinque 1990. Absence of quantifiers in Mohawk: Baker 1995, 1996.

109 Reference condition: Lasnik 1989; Reinhart 1983. Pronoun reference in Mohawk: Baker 1991, 1996.

111 The polysynthesis parameter, agreement-incorporation interactions: Baker 1996.

113 No infinitives in Mohawk: Baker 1996.

115 Other polysynthetic languages: Baker 1996.

116 The relatedness of Australian languages: Dixon 1980; Blake 1987.

117 Eskimo words for 'snow': Martin 1986; Pullum 1989; and personal communication from Eskimologists Jerry Sadock, Anthony Woodbury (1989), and Maria Bittner (1998).

118 The Sapir-Whorf hypothesis: Whorf 1956. Modern revivals: Gumprez and Levinson 1996. Traditional Mohawk culture: see, for example, Josephy 1991.

119 Noncorrelation of language type and society type: Hill 1993; Baker 1996.

120 Relative power of continents: Diamond 1997.

121 Acquisition of the head parameter: Slobin 1985.

CHAPTER 5

127 Word order in Welsh: Sproat 1985; Harlow 1981; King 1993. See also McCloskey 1991 for Modern Irish, a related language with similar properties.

127 Relative frequency of word orders: Tomlin 1986.

128 Warao as an object-subject-verb language: Romero-Figueroa 1985. For other candidates, see Derbyshire and Pullum 1981. Some of the object-subject-verb languages they cite (such as Urubú), however, have turned out to have a more normal subject-object-verb order upon further investigation.

130 Subject joined with verb phrase: Koopman and Sportiche 1991; McCloskey 1997.

132 Verb attraction to tense: Koopman 1984; Travis 1984. As a parameter: Chomsky 1995.

136 Adverbs and verb movement in French: Emonds 1978; Pollock 1989. All things being equal, one would expect a second kind of intermediate language as well: one in which the subject is attached to the verb phrase (as in Welsh) but the verb attracts the tense auxiliary (as in English). In fact, such languages do not exist, but explaining why requires a more sophisticated version of the verb attraction parameter than I can present here.

139 Warao as an object-subject-verb language: Romero-Figueroa 1985. Nadëb, a Tupian language spoken in Brazil, is another possible language of this type (Weir 1986).

141 Serial verb constructions: Foley and Olsen 1985; Sebba 1987; Baker 1989; Collins 1997.

142 Conflict between verb movement and serial verb constructions: Déchaine 1993; Baker and Stewart 1998.

142 Edo tense marking and serial verb constructions: Baker and Stewart 1998; Stewart 1998.

143 Serialization and word order in Khmer languages: Schiller 1990, 1991.

146 Chichewa word order and agreement: Bresnan and Mchombo 1987.

147 Slave word order and agreement: Rice 1989.

147 No languages with object agreement and not subject agreement: Croft 1990 and others.

153 Effects of optional object agreement: Givón 1976; Moravcsik 1974. The Swahili examples are from unpublished work by Niki Keach.

154 Subjects as animate and definite: Givón 1976. Tagalog and Malagasy are languages in which subjects must be definite; Southern Tiwa and to a large extent Japanese are languages in which subjects must be animate.

155 Infinitives in Chichewa: Bresnan and Mchombo 1987.

CHAPTER 6

158 Chemistry at the time of Mendeleyev: Atkins 1995; Ley 1968.

160 Characteristics of Mendeleyev's first table: Atkins 1995; Ley 1968; Brock 1992.

167 Survey of object-initial languages: Derbyshire and Pullum 1981. Indirect objects in Hixkaryana: Derbyshire 1985. Subject-final languages as a result of movement: Kayne 1995: 36.

168 Absence of auxiliary-subject-verb-object: this is my own discovery, based on an analysis of the raw data in Julien 2000.

168 Null subjects and verb movement: Alexiadou and Anagnostopoulou 1998.

169 Relationship of null subjects to polysynthesis: Barbosa 1993; Alexiadou and Anagnostopoulou 1998.

171 Indo-European charts can be found in many reference works. This one happens to be based on the *Random House Dictionary of the English Language* (unabridged, 2nd edition).

171 Working toward Proto-World: Greenberg 1987, 2000. Controversies about large-scale historical linguistics: Campbell 1988, 1997; Greenberg 1989.

176 Adjectives neutralized with other categories: Dixon 1982; Bhat 1994; among others. Mayali example: Evans 1991.

177 Split noun phrases in languages with adjectives like nouns: Bhat 1994; Baker 2001. (But I retreat from this position in Baker forthcoming.)

178 Case marking as typical of head-last languages: Greenberg 1963: 96.

179 Case in Japanese: Kuno 1973. Ergative case in Greenlandic: Bok-Bennema 1991; Bittner 1994; Bittner and Hale 1996b. On ergative languages in general: Dixon 1979, 1994.

180 Ergative languages have different grammatical relations: Dixon 1979, 1994. Ergative languages violate the verb-object constraint: Marantz 1984. Ergative language speakers are passive: discussion at a Wenner-Gren symposium, Ocho Rios, Jamaica, 1987.

180 The near equivalence of ergative and accusative languages: Anderson 1976; Bittner and Hale 1996a; and others. Noun incorporation as evidence of subjects and objects in ergative languages: Baker 1988.

183 Topic-prominent languages: Li and Thompson 1976. Topic in Japanese: Kuno 1973. Somali and Quechua as other topic prominent languages: Kiss 1995. Choctaw: Broadwell 1990.

184 Question movement parameter: Huang 1982; Lasnik and Saito 1984; and many others.

185 Question movement correlated with word order: Bach 1971. No question movement in Chichewa: Bresnan and Mchombo 1987. Question movement in Quechua: Lefebvre and Muysken 1988.

185 Question movement required in polysynthetic languages: Baker 1995, 1996.

187 Reflexives in Chinese: Huang 1982; Huang and Tang 1991. Explicit parameters for reflexives: Manzini and Wexler 1987.

189 No tense in Hopi: Whorf 1956. In Mohawk: Baker and Travis 1997. In Eskimo: Shaer 1992.

189 Parametric differences recast as lexical differences: Chomsky 1995. The idea has its origins in Borer 1984.

190 Lexical aspects of reflexive differences: Pica 1987; and many others. *Taziji* in Chinese: Huang and Tang 1991.

193 Early setting of head direction: Bloom 1970; Brown 1973; Slobin 1985.

193 Verb attraction in child English and French: Deprez and Pierce 1993.

194 Subject placement in child English and French: Deprez and Pierce 1993.

195 Null subjects in child English: Hyams 1989. See also Snyder and
     Stromswold 1997 for the acquisition of another parameter distin-
     guishing English and French, acquired around twenty-nine months.
195 Acquisition of ergativity and accusative case: Pye 1990.
196 Quantum mechanics explicating the periodic table: Atkins 1995:
     ch. 9.

CHAPTER 7

202 Sapir on limitless variation: Sapir 1921. Greenberg universals: Green-
     berg 1963.
203 Mohawk distantly related to Siouan: Chafe 1976. (Campbell 1997,
     however, evaluates the support for this view as fairly weak.) System-
     atic differences between them: Baker 1996.
204 Derrida's postmodernism: Derrida 1976. Derrida's views are expli-
     cated in many textbooks; Sarup 1993 is helpful because it has more on
     Derrida's relationship to Saussure than many sources.
206 Comparing human language to other species-specific biological abili-
     ties: Pinker and Bloom 1990; Pinker 1994.
207 Human language and elephant trunks: Pinker 1994. Evolution of lan-
     guage: Bickerton 1990, 1995; Carstairs-McCarthy 1999; Newmeyer
     1991; Lieberman 1985, 1999.
207 Silence of evolution on linguistic diversity: Lieberman 1985, 1999;
     Bickerton 1990, 1995; Carstairs-McCarthy 1999; Newmeyer 1991.
208 Evolution of parameters in restricted grammar: Nowak, Komarova, et
     al. 2001. Nowak investigates the population dynamic effects of vary-
     ing n, the number of languages that are consistent with an innate en-
     dowment. My *Homo whateverus* corresponds to a variety with n
     greater than a coherence threshold that he calculates, my *Homo para-
     meterus* to a variety with n smaller than the coherence threshold, and
     my *Homo rigidus* to a variety with n at the lower limit of 1.
208 Principles and parameters of bee navigation: Gallistel 1990, 1995.
209 Benefits of allowing some learning: Aitchison 1996.
210 Advantages of dividing humanity: Dyson 1979. Inadequacies of the
     view: Pinker 1994.
211 Selfish genes, altruism, and kin recognition: Dawkins 1976.
213 No motive to fully specify grammar: Pinker and Bloom 1990; Pinker
     1994.
213 Difficulties with learning first parameters: Gibson and Wexler 1994;
     Fodor 1998.

215 Genetic syndromes affecting language: Gopnik and Crago 1991.

215 Possible origins of language in primate conceptual systems: Bickerton 1990, 1995.

216 Nonadaptionist biological explanations of language: Chomsky 1972, 1988; Piattelli-Palmarini 1989.

216 Biology refuting principles and parameters linguistics: Lieberman 1999.

218 Language diversification compared to speciation: Pinker 1994; Darwin 1874; and others.

218 Stages of language change: Pinker 1994; Lightfoot 1999; among many others.

220 Word order change in English: Lightfoot 1991.

222 Passive leading to ergativity in Polynesian languages: Chung 1978.

223 Unpredictable use of language and free will: Chomsky 1966. Cartesian roots: Descartes 1996.

224 Language not learned from TV: Sachs, Bard, et al. 1981; Levelt 1989; Pinker 1994.

226 Abruptness of word order change in English: Lightfoot 1991.

228 Puzzles and mysteries: Chomsky 1975; McGinn 1993; Pinker 1997.

230 Rapid extinction of languages: Hale, Krauss, et al. 1992.

233 Different dimensions of verb meanings in different languages: Talmy 1985.

233 Possible consequences of speaking different languages: Gumprez and Levinson 1996.

Glossary of
Linguistic Terms

Accusative case: A prefix, suffix, or particle that is attached to a noun phrase or a pronoun to indicate that it is the direct object of the clause.

Adjective: A major-class word that is not a noun or a verb; it typically describes a quality or state. Examples: *tall, hungry, transparent*.

Adverb: An optional word or phrase that is added to a sentence to give additional information about the event being described. Examples: *quickly, later, in the park*.

Agent: The person or thing responsible for initiating the event described by the verb.

Agreement: A prefix or suffix that is added to a verb (or other word) to match the person, number, and/or gender of a designated noun phrase that is syntactically related to that verb (generally the subject or the object). Example: *I think* vs. *Kate thinks*. Agreement is typical of head-marking languages.

Antecedent: A noun phrase that refers to the same person or thing as a subsequent pronoun or reflexive pronoun. Example: In *Julia lost her dog*, *Julia* is the antecedent of *her*.

Article: A minor-category word that appears together with a noun phrase in some languages. Often it expresses whether the noun phrase has a definite, specific, or indefinite reference. Examples: *a book, the cat*.

Auxiliary: A minor-category word or extra verb that appears together with a verb phrase in some languages, often expressing tense or mood. Examples: *I will come; Nicholas might win*.

Case marker: A prefix, suffix, or particle that is attached to a noun phrase or pronoun to indicate the function of that noun phrase within the clause as a whole. Case markers are typical of dependent-marking languages.

Causative: A prefix or suffix that attaches to a root to add the idea that the state of affairs was caused by something or someone specific. Example: *I enlarged the picture* expresses not only that the picture is large but that I caused it to be so.

Clause: A basic unit of grammatical structure, expressing a single thought. A clause typically consists of one subject (noun phrase) and one predicate (verb phrase), together with any related adverbs.

Complementizer: A minor-category word, a type of conjunction, that marks one clause as being embedded in another. Example: *Julia thinks that it will rain.*

Completeness condition: The condition that all the basic participants of an event must be expressed somehow in the clause that describes that event.

Conjunction: A minor category of words that serves to connect two phrases into a larger phrase (example: *two cats and a dog*). Complementizers are a subclass of conjunctions.

Copular verb: A verb with little inherent meaning that connects a subject to some kind of predicate, especially a predicate that is not a verb. Example: *Kate is hungry.*

Dative case: A prefix, suffix, or particle that is attached to a noun phrase or a pronoun to indicate that it is the indirect object of the clause.

Dependent: Any member of a phrase that is not the head of the phrase. For example, subjects, objects, indirect objects, and adverbs are all dependents of the verb inside a clause. The possessor is a dependent of the noun inside a noun phrase.

Direct object: A noun phrase that is directly related to the verb, with no preposition or postposition. In simple sentences it typically expresses the undergoer of the event described. Example: *I sent a letter to Nicholas.*

Dislocation: The shifting of a contentful noun phrase from its ordinary position inside a clause to the edge of the clause, leaving a pronoun or affix behind. Example: *That song, I really liked it,* as compared to *I really liked that song.*

E-language: "External" language; language seen as a set of sentences and other structures, considered independently of the mind.

Embedded clause: A clause that is contained inside a larger clause. Example: *Julia thinks that it will rain.*

Ergative: A case marker that attaches to the subject noun phrase only in transitive sentences. The term may also apply to a language that has such a case marker.

Gender: A division of nouns into smaller classes, such as masculine, feminine, and neuter. Often these nouns of different classes trigger different agreement affixes on the associated verb or adjective.

Grammar: The part of a language responsible for assembling basic words into larger words, phrases, and clauses in systematic ways. For purposes of this book, grammar is roughly the same as the syntax of a language plus its morphology.

Head: The word that a phrase is built around. For example, *give* is the head of the verb phrase *give a book to Kate; in* is the head of the prepositional phrase *in the park; rumors* is the head of the noun phrase *rumors about Nicholas.*

I-language: "Internal" language; language seen as a system of rules and principles in the human mind; language as a recipe.

Implicational universal: A general statement about human languages of the form that all languages that have a certain feature A also have a second feature B. Example: all languages that have the sound *b* also have the sound *p.*

Incorporation: The process by which two expressions that could otherwise be separate words are fused into a single word in some languages— particularly the noun expressing the direct object being fused with the verb in polysynthetic languages.

Indirect object: The noun phrase that expresses the goal, endpoint, or beneficiary of the event expressed by the verb of the clause. Example: *I sent a letter <u>to Nicholas.</u>*

Infinitive: A clause in which the verb does not take its usual marking for tense and agreement. Example: *I saw Julia leave* (not **I saw Julia leaves* or **I saw Julia left*).

Inflection: The affixes that attach to a word to make it suitable for a particular grammatical context. Tense and agreement are common verb inflections; number and case are common noun inflections.

Intransitive: A clause or verb that has a subject but no direct object.

Lexicon: The stock of basic words of a language.

Mood: Similar to tense, mood expresses whether an event is possible, certain, or impossible, whether it is required or forbidden. Example: *Kate <u>might</u> win.*

Morphology: The branch of linguistics that studies how words are formed from roots and stems, prefixes and suffixes.

Nominative case: A prefix, suffix, or particle that is attached to a noun phrase or a pronoun to indicate that it is the subject of the clause.

Nonconfigurational language: A language in which words are freely ordered, noun phrases are often omitted, and little phrase formation seems to take place.

Noun: A major category of words that have the capacity to refer to a person or thing. Examples: *dog, bachelor, water, liberty.*

Number: An inflection on nouns and pronouns, roughly indicating whether they refer to one thing (singular) or to more than one thing (plural).

Object: A noun phrase that functions as a dependent in a phrase with some other head. See *direct object, indirect object.* Prepositional phrases can also contain objects, as in *on the table*. Unless otherwise specified, I use *object* as short for *direct object*.

Parameter: The atoms of linguistic diversity; a choice point in a recipe for language, permitted by universal grammar. Different choices about how to do things at this point lead to different types of language.

Particle: Any minor-category word that does not take inflections. Often a minor grammatical element that linguists don't feel like figuring out.

Passive voice: A rearranged sentence in which the undergoer is expressed as the subject (instead of the direct object) and the agent is either omitted or expressed as some kind of prepositional phrase. Example: *Chris was killed* or *Chris was killed by a falling tree,* as opposed to the active sentence *A falling tree killed Chris.*

Person: A classification of nouns and pronouns depending on whether they refer to the speaker (first person: *I, we*), the hearer (second person: *you*), or someone else (third person: *he, she, it, they, Chris,* etc.).

Phonology: The branch of linguistics that studies sounds and how they affect one another.

Phrase: A group of words that appear next to each other, that tend to stay together in rearrangements of a sentence, and that form a semantic unit.

Polysynthetic language: A language with long, complex verbs, in which many grammatical relationships are expressed by affixes. Polysynthetic languages have lots of agreement and incorporation.

Possessor: A noun phrase that is contained within a larger noun phrase, expressing a possessive relation with the head noun. Example: <u>*Nicholas's*</u> *cat.*

Postposition: The same category as a preposition, but in a language where heads come at the end of phrases.

Predicate: The phrase that is combined with a subject to form a clause. The predicate is typically a verb phrase or an auxiliary phrase. Example: *Julia* <u>*will leave soon.*</u>

Prefix: A meaningful element that cannot stand on its own but is added to the beginning of another element (a stem). Example: <u>*re*</u>*think,* <u>*in*</u>*visible,* <u>*anti*</u>*matter.*

Preposition: A class of words that typically go before noun phrases and create phrases that can be used as adverbs or modifiers. Often they express spatial or temporal relations. Examples: <u>*in*</u> *the park,* <u>*since*</u> *last week.*

Pronoun: A minor class of words that take the place of a noun phrase. Typically its reference can be deduced from an antecedent or from some other feature of the context. Examples: *he, she, it, they, we.*

Quantifier, nonreferential: A noun-phrase-like construction that does not refer but rather quantifies how many things satisfy the description. Examples: <u>*Nobody*</u> *came;* <u>*Everyone*</u> *likes Julia.*

Reference: The capacity of nouns and pronouns to designate a particular entity. Also, the entity designated by a given noun phrase or pronoun.

Reflexive pronoun: A special class of pronouns used to indicate (for example) that two participants in an event are the same person or thing. Example: *Kate embarrassed herself at the party.*

Root: The core of a word, before any prefixes or suffixes have been attached. Example: the root of *antidisestablishmentarianism* is *establish*.

Semantics: The branch of linguistics that is concerned with the meanings of linguistic expressions.

Sentence: A basic unit of language, composed of words and expressing a complete thought. A clause that is not embedded.

Stem: That which a prefix or suffix attaches to in order to form a larger word. The affix *-ism* attaches to the stem *antidisestablishmentarian* to form the word *antidisestablishmentarianism*.

Subject: A noun phrase that combines with the predicate phrase to form a clause. In simple sentences it typically expresses the agent of the event described. Example: *I sent a letter to Nicholas.*

Suffix: A meaningful element that cannot stand on its own but is added to the end of another element (a stem). Examples: *legalize, breakable, cooking.*

Syntax: The branch of linguistics that studies how words are combined to make phrases and sentences.

Tense: A word or affix that expresses the time value of the clause, whether the event described by the verb takes place in the past, present, or future.

Topic phrase: A phrase (typically a noun phrase) that is added to the edge of a clause and expresses what the clause is about. Example: *As for this country, people generally live a long time here.*

Transitive: A clause or verb that has a direct object as well as a subject.

Undergoer: The participant of an event that is most changed as a result of the event.

Universal grammar: The innate knowledge that humans have about language, which, under the right circumstances, enables them to acquire any particular language spoken around them.

Verb: A major class of words that provide the core that a sentence can be built out of by the addition of noun phrases and other elements. Verbs typically describe events that involve participants or states that hold of things. Examples: *run, hit, give, shine.*

Word: A freestanding chunk of language with a coherent meaning. The building blocks out of which phrases and sentences can be constructed.

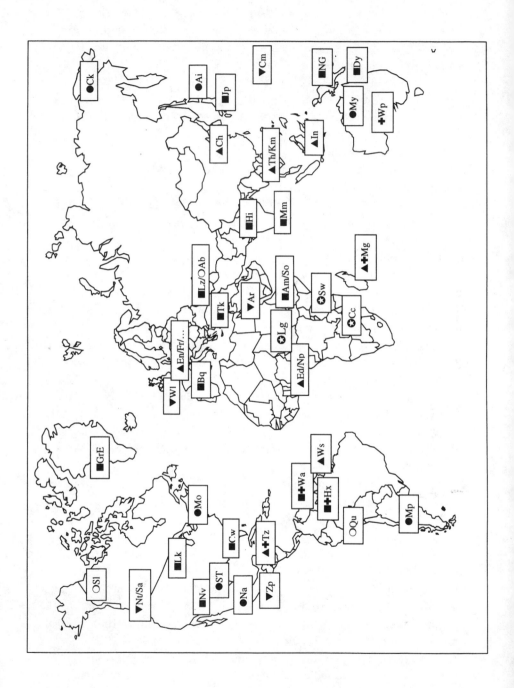

Map of the world, showing approximately where the principal languages discussed are spoken. Notice that grammatically similar languages can be spoken in very different parts of the world.

Head-Last Languages

■ Simple subject-object-verb

Am Amharic (Afro-Asiatic)
Bq Basque
Cw Choctaw (Muskogean)
Dy Dyirbal (Australian)
GrE Greenlandic Eskimo
Hi Hindi (Indo-European)
Jp Japanese
Lk Lakhota (Macro-Siouan)
Lz Lezgian (Caucasian)
Mm Malayalam (Dravidian)
NG New Guinean languages (various)
Nv Navajo (Athapaskan)
So Somali (Afro-Asiatic)
Tk Turkish (Altaic)

■+ Object-verb-subject

Hx Hixkaryana (Carib)
Wa Warao

Head-First Languages

▲ Simple subject-verb-object

Ch Chinese (Sino-Tibetan)
Ed Edo (Niger-Congo)
En English (Indo-European)
Fr French (Indo-European)
In Indonesian (Austronesian)
Km Khmer (Mon-Khmer)
Np Nupe (Niger-Congo)
Th Thai
Ws Wapishana (Arawakan)

▼ Verb-subject-object

Ar Arabic (Afro-Asiatic)
Cm Chamorro (Austronesian)
Nt Nootka (Wakashan)
Sa Salish
Wl Welsh (Indo-European)
Zp Zapotec (Oto-Manguean)

▲+ Verb-object-subject

Mg Malagasy (Austronesian)
Tz Tzotzil (Mayan)

Partly Polysynthetic Languages

● Polysynthetic

Ai Ainu
Ck Chuckchee (Paleo-Siberian)
Mo Mohawk (Macro-Siouan)
Mp Mapuche
My Mayali (Australian)
Na Nahuatl (Uto-Aztecan)
ST Southern Tiwa (Tanoan)

○ SOV and polysynthesis

Ab Abkhaz (Caucasian)
Qu Quechua
Sl Slave (Athapaskan)

✪ SVO and polysynthesis

Cc Chichewa (Niger-Congo)
Lg Lango (Nilo-Saharan)
Sw Swahili (Niger-Congo)

+ Other Nonconfigurational

Wp Warlpiri (Australian)

References

Abney, Steven. 1987. *The English noun phrase in its sentential aspect.* Ph.D. dissertation, MIT.

Agheyisi, Rebecca. 1990. *A grammar of Edo.* Manuscript, UNESCO.

Aissen, Judith. 1987. *Tzotzil clause structure.* Dordrecht: Reidel.

Aitchison, Jean. 1996. *The seeds of speech: language origin and evolution.* Cambridge: Cambridge University Press.

Alexiadou, Artemis, and Elena Anagnostopoulou. 1998. Parametrizing AGR: word order, V-movement and EPP-checking. *Natural Language and Linguistic Theory* 16:491–539.

Amberber, Mengistu. 1997. *Transitivity alternations, event-types, and light verbs.* Ph.D. dissertation, McGill University.

Anderson, Stephen. 1976. On the notion of subject in ergative languages. In *Subject and topic,* ed. Charles Li, 1–23. New York: Academic Press.

Atkins, P. W. 1995. *The Periodic Kingdom.* New York: Basic Books.

Bach, Emond. 1971. Questions. *Linguistic Inquiry* 2:153–166.

Baker, Mark. 1988. *Incorporation: a theory of grammatical function changing.* Chicago: University of Chicago Press.

———. 1989. Object sharing and projection in serial verb constructions. *Linguistic Inquiry* 20:513–553.

———. 1991. On some subject/object non-asymmetries in Mohawk. *Natural Language and Linguistic Theory* 9:537–576.

———. 1995. On the absence of certain quantifiers in Mohawk. In *Quantification in natural languages,* ed. Emond Bach, Eloise Jelinek, Angelika Kratzer, and Barbara Partee, 21–58. Dordrecht: Kluwer.

———. 1996. *The polysynthesis parameter.* New York: Oxford University Press.

———. 1997. Thematic roles and syntactic structure. In *Elements of grammar,* ed. Liliane Haegeman, 73–137. Dordrecht: Kluwer.

———. 2001. The natures of nonconfigurationality. In *The handbook of syntax,* ed. Mark Baltin and Chris Collins, 407–438. Cambridge: Blackwell.

_____. Forthcoming. *On the lexical category distinctions.* Cambridge: Cambridge University Press.

Baker, Mark, and O. T. Stewart. 1998. Verb movement, objects, and serialization. In *Proceedings of the North East Linguistics Society* 29:17–32.

Baker, Mark, and Lisa Travis. 1997. Mood as verbal definiteness in a "tenseless" language. *Natural Language Semantics* 5:213–269.

Barbosa, Pilar. 1993. Clitic placement in Old Romance and European Portuguese and the null subject parameter. Manuscript, MIT.

Bernstein, Alison. 1991. *American Indians and World War II.* Norman: University of Oklahoma Press.

Bhat, D. N. S. 1994. *The adjectival category: criteria for differentiation and identification.* Amsterdam: John Benjamins.

Bickerton, Derek. 1990. *Language and species.* Chicago: University of Chicago Press.

_____. 1995. *Language and human behavior.* Seattle: University of Washington Press.

Bittner, Maria. 1994. *Case, scope, and binding.* Dordrecht: Kluwer.

Bittner, Maria, and Kenneth Hale. 1996a. Ergativity: toward a theory of a heterogeneous class. *Linguistic Inquiry* 27:531–604.

_____. 1996b. The structural determination of case and agreement. *Linguistic Inquiry* 27:1–68.

Blake, Barry. 1987. *Australian aboriginal grammar.* London: Croom Helm.

Bloom, Lois. 1970. *Language development: form and function in emerging grammars.* Cambridge: MIT Press.

Bloomfield, Leonard. 1970. *A Leonard Bloomfield anthology.* Bloomington: University of Indiana Press.

Bok-Bennema, Reineke. 1991. *Case and agreement in Inuit.* Berlin: Foris.

Borer, Hagit. 1984. *Parametric syntax: case studies in Semitic and Romance languages.* Dordrecht: Foris.

Bresnan, Joan, ed. 1982. *The mental representation of grammatical relations.* Cambridge: MIT Press.

Bresnan, Joan, and Sam Mchombo. 1987. Topic, pronoun, and agreement in Chichewa. *Language* 63:741–782.

Broadwell, George A. 1990. *Extending the binding theory: a Muskogean case study.* Ph.D. dissertation, University of California–Los Angeles.

Brock, William. 1992. *The Norton history of chemistry.* New York: W. W. Norton.

Brown, Roger. 1973. *A first language: the early stages.* Cambridge: Harvard University Press.

Campbell, Lyle. 1988. Review article: Greenberg, *Language in the Americas. Language* 64:591–615.

_____. 1997. *American Indian languages: the historical linguistics of Native America.* New York: Oxford University Press.

_____. 2001. The history of linguistics. In *The handbook of linguistics,* ed. Mark Aronoff and Janie Rees-Miller, 81–104. Oxford: Blackwell.

Carstairs-McCarthy, Andrew. 1999. *The origins of complex language.* Oxford: Oxford University Press.

Chafe, Wallace. 1976. Siouan, Iroquoian, and Caddoan. In *Native languages of the Americas,* ed. Thomas Sebeok, 1164–1209. New York: Plenum Press.

Chomsky, Noam. 1966. *Cartesian linguistics.* New York: Harper and Row.

_____. 1972. *Language and mind.* New York: Harcourt Brace Jovanovich.

_____. 1975. *Reflections on language.* New York: Pantheon.

_____. 1980. *Rules and representations.* New York: Columbia University Press.

_____. 1981. *Lectures on government and binding.* Dordrecht: Foris.

_____. 1986a. *Barriers.* Cambridge: MIT Press.

_____. 1986b. *Knowledge of language: its nature, origin, and use.* New York: Praeger.

_____. 1988. *Language and problems of knowledge.* Cambridge: MIT Press.

_____. 1995. *The minimalist program.* Cambridge: MIT Press.

Chung, Sandra. 1978. *Case marking and grammatical relations in Polynesian.* Austin: University of Texas Press.

Cinque, Guglielmo. 1990. *Types of A-bar dependencies.* Cambridge: MIT Press.

Collins, Chris. 1997. Argument sharing in serial verb constructions. *Linguistic Inquiry* 28:461–497.

Croft, William. 1990. *Typology and universals.* Cambridge: Cambridge University Press.

_____. 1995. Autonomy and functionalist linguistics. *Language* 71:490–532.

Darwin, Charles. 1874. *The descent of man and selection in relation to sex.* New York: Hurst & Co.

Dawkins, Richard. 1976. *The selfish gene.* New York: Oxford University Press.

Déchaine, Rose-Marie. 1993. *Predicates across categories.* Ph.D. dissertation, University of Massachusetts–Amherst.

Deering, Nora, and Helga Delisle. 1976. *Mohawk: a teaching grammar.* Kahnawake, Quebec: Thunderbird Press.

Deprez, Viviane, and Amy Pierce. 1993. Negation and functional projections in early child grammar. *Linguistic Inquiry* 24:25–68.

Derbyshire, D. C. 1985. *Hixkaryana and linguistic typology.* Arlington, Tex.: Summer Institute of Linguistics.

Derbyshire, Desmond, and Geoffrey Pullum. 1981. Object initial languages. *International Journal of American Linguistics* 47:192–214.

Derrida, Jacques. 1976. *Of grammatology.* Baltimore: Johns Hopkins University Press.

Descartes, Rene. 1996. *Discourse on the method and meditations on first philosophy.* Ed. David Weissman. New Haven, Conn.: Yale University Press.

Diamond, Jared. 1997. *Guns, germs, and steel.* New York: W. W. Norton.

Dixon, R. M. W. 1979. Ergativity. *Language* 55:59–138.

_____. 1980. *The languages of Australia.* Cambridge: Cambridge University Press.

_____. 1982. *Where have all the adjectives gone?* Berlin: Mouton de Gruyter.

_____. 1994. *Ergativity.* Cambridge: Cambridge University Press.

Dobrovie-Sorin, Carmen. 1994. *The syntax of Romanian.* Berlin: Mouton de Gruyter.

Dryer, Matthew. 1992. The Greenbergian word order correlations. *Language* 68:81–138.

Dyson, Freeman. 1979. *Disturbing the universe.* New York: Harper.

Emonds, Joseph. 1978. The verbal complex V'-V in French. *Linguistic Inquiry* 9:151–175.

Evans, Nicholas. 1991. A draft grammar of Mayali. Manuscript, University of Melbourne.

Fodor, Janet Dean. 1998. Unambiguous triggers. *Linguistic Inquiry* 29:1–37.

Foley, William. 1991. *The Yimas language of New Guinea.* Stanford: Stanford University Press.

Foley, William, and Michael Olsen. 1985. Clausehood and verb serialization. In *Grammar inside and outside the clause,* ed. Johanna Nichols and Anthony Woodbury, 17–60. Cambridge: Cambridge University Press.

Foley, William, and Robert Van Valin. 1984. *Functional syntax and universal grammar.* Cambridge: Cambridge University Press.

Fukui, Naoki, and Margaret Speas. 1986. Specifiers and projections. MIT *Working Papers in Linguistics* 8:128–172.

Gaarder, Jostein. 1994. *Sophie's world.* New York: Berkley Books.

Gallistel, C. R. 1990. *The organization of learning.* Cambridge: MIT Press.

_____. 1995. Symbolic processes in the brain: the case of insect navigation. In *An invitation to cognitive science,* ed. Edward Smith and Daniel Osherson, 1–51. Cambridge: MIT Press.

Gibson, Edward, and Kenneth Wexler. 1994. Triggers. *Linguistic Inquiry* 25:407–454.

Gilligan, Gary. 1987. A crosslinguistic approach to the pro-drop parameter. Ph.D. dissertation, University of California–Los Angeles.

Givón, Talmy. 1976. Topic, pronoun, and grammatical agreement. In *Subject and topic,* ed. Charles Li, 149–189. New York: Academic Press.

———. 1979. *On understanding grammar.* New York: Academic Press.

Gopnik, Myrna, and Martha Crago. 1991. Familial aggregation of a developmental language disorder. *Cognition* 39:1–50.

Greenberg, Joseph. 1963. *Universals of language.* Cambridge: MIT Press.

———. 1987. *Language in the Americas.* Stanford: Stanford University Press.

———. 1989. Classification of American Indian languages: a reply to Campbell. *Language* 65:107–114.

———. 2000. *Indo-European and its closest relatives: the Eurasiatic language family.* Stanford: Stanford University Press.

Grimes, Barbara, ed. 1988. *Ethnologue.* Dallas, Tex.: Summer Institute of Linguistics.

Grimshaw, Jane. 1997. Projections, heads, and optimality. *Linguistic Inquiry* 28:373–422.

Guilfoyle, Eithne, Henrietta Hung, and Lisa Travis. 1992. Spec of IP and Spec of VP: two subjects in Austronesian languages. *Natural Language and Linguistic Theory* 10:375–414.

Gumprez, John, and Stephen Levinson, ed. 1996. *Rethinking linguistic relativity.* Cambridge: Cambridge University Press.

Hale, Kenneth. 1973. A note on subject-object inversion in Navajo. In *Issues in linguistics: papers in honor of Henry and Renee Kahane,* ed. B. Kachru, 300–309. Chicago: University of Illinois Press.

———. 1983. Warlpiri and the grammar of nonconfigurational languages. *Natural Language and Linguistic Theory* 1:5–49.

Hale, Kenneth, Michael Krauss, Lucille Watahomigie, Akira Yamamoto, Colette Craig, Laverne Masayesva Jeanne, and Nora England. 1992. Endangered languages. *Language* 68:1–42.

Harlow, Stephen. 1981. Government and relativization in Celtic. In *Binding and filtering,* ed. Frank Heny, 213–254. Cambridge: MIT Press.

Hawkins, John. 1990. A parsing theory of word order universals. *Linguistic Inquiry* 21:223–261.

———. 1994. *A performance theory of order and constituency.* Cambridge: Cambridge University Press.

Hill, Jane. 1993. Formalism, functionalism, and the discourse of evolution. In *The role of theory in language description,* ed. William Foley, 437–455. Berlin: Mouton de Gruyter.

Huang, C.-T. James. 1982. *Logical relations in Chinese and the theory of grammar*. Ph.D. dissertation, MIT.

Huang, C.-T. James, and C.-C. Jane Tang. 1991. The local nature of the long-distance reflexive in Chinese. In *Long-distance anaphora,* ed. Jan Koster and Eric Reuland, 263–282. Cambridge: Cambridge University Press.

Hyams, Nina. 1989. The null subject parameter in language acquisition. In *The null subject parameter,* ed. Osvaldo Jaeggli and Kenneth Safir, 215–238. Dordrecht: Kluwer.

Jaeggli, Osvaldo. 1982. *Topics in Romance syntax*. Dordrecht: Foris.

Jaeggli, Osvaldo, and Kenneth Safir. 1989. The null subject parameter and parametric theory. In *The null subject parameter,* ed. Osvaldo Jaeggli and Kenneth Safir, 1–44. Dordrecht: Kluwer.

Jelinek, Eloise. 1984. Empty categories, case, and configurationality. *Natural Language and Linguistic Theory* 2:39–76.

Jelinek, Eloise, Sally Midgette, Keren Rice, and Leslie Saxon, ed. 1996. *Athabaskan language studies: essays in honor of Robert W. Young*. Albuquerque: University of New Mexico Press.

Johnson, Broderick, ed. 1977. *Navajos and World War II*. Tsaile, Navajo Nation, Ariz.: Navajo Community College Press.

Josephy, Alvin, ed. 1991. *America in 1492*. New York: Random House.

Julien, Marit. 2000. Syntactic heads and word formation. Ph.D. dissertation, University of Tromsoe, Norway.

Kayne, Richard. 1984a. *Connectedness and binary branching*. Dordrecht: Foris.

_____. 1984b. Principles of particle constructions. In *Grammatical representation,* ed. Jacqueline Guéron, Hans-Georg Obenauer, and Jean-Yves Pollock, 101–140. Dordrecht: Foris.

_____. 1995. *The antisymmetry of syntax*. Cambridge: MIT Press.

Keenan, Ed. 1976. Remarkable subjects in Malagasy. In *Subject and topic,* ed. C. N. Li. New York: Academic Press.

King, Gareth. 1993. *Modern Welsh: a comprehensive grammar*. London: Routledge.

Kiss, Katalin, ed. 1995. *Discourse configurational languages*. New York: Oxford University Press.

Koopman, Hilda. 1984. *The syntax of verbs*. Dordrecht: Foris.

Koopman, Hilda, and Dominique Sportiche. 1991. The position of subjects. *Lingua* 85:211–258.

Kuno, Susumu. 1973. *The structure of the Japanese language*. Cambridge: MIT Press.

Lasnik, Howard. 1989. *Essays on anaphora*. Dordrecht: Kluwer.

Lasnik, Howard, and Mamoru Saito. 1984. On the nature of proper government. *Linguistic Inquiry* 15:235–299.

Lefebvre, Claire, and Pieter Muysken. 1988. *Mixed categories: nominalizations in Quechua.* Dordrecht: Kluwer.

Levelt, Willem. 1989. *Speaking: from intention to articulation.* Cambridge: MIT Press.

Ley, Willy. 1968. *The discovery of the elements.* New York: Delacorte Press.

Li, Charles, and Sandra Thompson. 1976. Subject and topic: a new typology. In *Subject and topic,* ed. Charles Li, 457–489. New York: Academic Press.

Lieber, Rocelle. 1983. Argument linking and compounding in English. *Linguistic Inquiry* 14:251–286.

Lieberman, Philip. 1985. *The biology and evolution of language.* Cambridge: Harvard University Press.

_____. 1999. *Eve spoke: human language and human evolution.* New York: W. W. Norton.

Lightfoot, David. 1991. *How to set parameters.* Cambridge: MIT Press.

_____. 1999. *The development of language: acquisition, change, and evolution.* Malden, Mass.: Blackwell.

Lord, Carol. 1993. *Historical change in serial verb constructions.* Amsterdam: John Benjamins.

Lounsbury, Floyd. 1953. *Oneida verb morphology.* New Haven, Conn.: Yale University Press.

Manzini, Rita, and Kenneth Wexler. 1987. Parameters, binding theory, and learnability. *Linguistic Inquiry* 18:413–444.

Marantz, Alec. 1984. *On the nature of grammatical relations.* Cambridge: MIT Press.

Martin, Laura. 1986. Eskimo words for snow: a case study in the genesis and decay of an anthropological example. *American Anthropologist* 88:418–423.

McCloskey, James. 1991. Clause structure, ellipsis and proper government in Irish. *Lingua* 85:259–302.

_____. 1997. Subjecthood and subject positions. In *Elements of grammar,* ed. Liliane Haegeman, 197–236. Dordrecht: Kluwer.

McGinn, Colin. 1993. *Problems in philosophy: the limits of inquiry.* Cambridge: Blackwell.

Mithun, Marianne. 1984. The evolution of noun incorporation. *Language* 60:847–893.

_____. 1987. Is basic word order universal? In *Coherence and grounding in discourse,* ed. R. Tomlin, 281–328. Amsterdam: John Benjamins.

Mithun, Marianne, and Wallace Chafe. 1979. Recapturing the Mohawk language. In *Languages and their status,* ed. Timothy Shopen, 3–34. Cambridge, Mass.: Winthrop.

Moravcsik, Edith. 1974. Object-verb agreement. *Working Papers in Language Universals* 15:25–140.

Newmeyer, Frederick. 1991. Functional explanation in linguistics and the origins of language. *Language and Communication* 11:3–114.

_____. 1998. *Language form and language function.* Cambridge: MIT Press.

Newquist, Harvey. 1994. *The brain makers.* Indianapolis: Sams Publishing.

Nichols, Johanna. 1986. Head-marking and dependent-marking grammar. *Language* 62:56–119.

_____. 1992. *Linguistic diversity in space and time.* Chicago: University of Chicago Press.

Nowak, Martin, Natalia Komarova, and Partha Niyogi. 2001. Evolution of universal grammar. *Science* 291:114–118.

Paul, Doris. 1973. *The Navajo Code Talkers.* Philadelphia: Dorrance & Company.

Perlmutter, David. 1971. *Deep and surface structure constraints in syntax.* New York: Holt, Rinehart, and Winston.

Piattelli-Palmarini, M. 1989. Evolution, selection, and cognition. *Cognition* 31:1–44.

Pica, Pierre. 1987. On the nature of the reflexivization cycle. *Proceedings of the North East Linguistics Society* 17:483–500.

Pinker, Steven. 1994. *The language instinct.* New York: William Morrow and Company.

_____. 1997. *How the mind works.* New York: W. W. Norton.

Pinker, Steven, and Paul Bloom. 1990. Natural language and natural selection. *Behavioral and Brain Sciences* 13:707–784.

Pollock, Jean-Yves. 1989. Verb movement, universal grammar, and the structure of IP. *Linguistic Inquiry* 20:365–424.

Postal, Paul. 1979. *Some syntactic rules of Mohawk.* New York: Garland.

Prince, Alan, and Paul Smolensky. 1993. Optimality theory: constraint interaction in generative grammar. Rutgers Center for Cognitive Science Technical Report 2, Rutgers University.

Pullum, Geoffrey. 1989. The great Eskimo vocabulary hoax. *Natural Language and Linguistic Theory* 7:275–282.

Pye, Clifton. 1990. The acquisition of ergative languages. *Linguistics* 28, 6:1291–1330.

Rackowski, Andrea, and Lisa Travis. 2000. V-initial languages: X or XP movement and adverbial placement. Manuscript, McGill University.

Radford, Andrew. 1981. *Transformational syntax: a student's guide to Chomsky's Extended Standard Theory.* Cambridge: Cambridge University Press.

Reinhart, Tanya. 1983. *Anaphora and semantic interpretation.* Chicago: University of Chicago Press.

Reinhart, Tanya, and Eric Reuland. 1993. Reflexivity. *Linguistic Inquiry* 24:657–720.

Rice, Keren. 1989. *A grammar of Slave.* Berlin: Mouton de Gruyter.

Rizzi, Luigi. 1982. *Issues in Italian syntax.* Dordrecht: Foris.

_____. 1986. On the status of subject clitics in Romance. In *Studies in Romance linguistics,* ed. Osvaldo Jaeggli and C. Silva-Corvalan, 391–420. Dordrecht: Foris.

Robins, R. H. 1989. *A short history of linguistics.* London: Longman.

Roeper, Tom, and M. E. A. Siegel. 1978. A lexical transformation for verbal compounds. *Linguistic Inquiry* 9:199–260.

Romero-Figueroa, Andres. 1985. OSV as the basic word order in Warao. *Lingua* 66:115–134.

Sachs, J., B. Bard, and M. L. Johnson. 1981. Language learning with restricted input: case studies of two hearing children of deaf parents. *Journal of Applied Psycholinguistics* 2:33–54.

Sapir, Edward. 1921. *Language.* New York: Harcourt, Brace & Company.

Sapir, Edward, and Morris Swadesh. 1946. American Indian grammatical categories. *Word* 2: 103–112.

Sarup, Madan. 1993. *An Introductory Guide to Post-Structuralism and Post-Modernism.* Athens: University of Georgia Press.

Schiller, Eric. 1990. The typology of serial verb constructions. In *Papers from the 26th regional meeting of the Chicago Linguistics Society,* ed. Michael Ziolkowski, Manuela Noske, and Karen Deaton, 393–406. Chicago: Chicago Linguistics Society.

_____. 1991. *An autolexical account of subordinating serial verb constructions.* Ph.D. dissertation, University of Chicago.

Sebba, Mark. 1987. *The syntax of serial verbs.* Amsterdam: John Benjamins.

Selkirk, Elisabeth. 1982. *The syntax of words.* Cambridge: MIT Press.

Shaer, Benjamin. 1992. Tense, time reference, and the Eskimo verb. Manuscript, McGill University.

Slobin, Dan, ed. 1985. *The crosslinguistic study of language acquisition.* Hillsdale, N.J.: L. Erlbaum Associates.

Snyder, William, and Karin Stromswold. 1997. The structure and acquisition of English dative constructions. *Linguistic Inquiry* 28:281–317.

Speas, Margaret. 1990. *Phrase structure in natural language*. Dordrecht: Kluwer.

Sproat, Richard. 1985. Welsh syntax and VSO structure. *Natural Language and Linguistic Theory* 3:173–216.

Stewart, O. T. 1998. *The serial verb construction parameter*. Ph.D. dissertation, McGill University.

Stocking, George, Jr. 1974. The Boas plan for the study of American Indian Languages. In *Studies in the history of linguistics: traditions and paradigms*, ed. Dell Hymes, 454–486. Bloomington: University of Indiana Press.

Stowell, Timothy. 1981. Origins of phrase structure. Ph.D. dissertation, MIT.

Talmy, Leonard. 1985. Lexicalization patterns. In *Language typology and syntactic description*, ed. Timothy Shopen, 57–149. Cambridge: Cambridge University Press.

Tomlin, Russell. 1986. *Basic word order: functional principles*. London: Croom Helm.

Travis, Lisa. 1984. *Parameters and effects of word order variation*. Ph.D dissertation, MIT.

Turing, A. M. 1950. Computing machinery and intelligence. *Mind* 65:433–460.

Weir, E. M. Helen. 1986. Footprints of yesterday's syntax: diachronic development of certain verb prefixes in an OSV language (Nadeb). *Lingua* 68:291–316.

Whaley, Lindsay. 1997. *Introduction to typology*. Thousand Oaks, Calif.: Sage.

Whorf, Benjamin. 1956. *Language, thought, and reality*. Cambridge: MIT Press.

Williamson, Janis. 1984. *Studies in Lakhota grammar*. Ph.D. dissertation, University of California–San Diego.

Young, Robert, and Stanley Morgan. 1987. *The Navajo language: a grammar and colloquial dictionary*. Albuquerque: University of New Mexico Press.

Zepter, Alex. 2000. Mixed word order: left or right, that is the question. Manuscript, Rutgers University.

Index

in polysynthetic language, 88–89
question making and, 37
reference condition and, 109–110
reflexive sentences and, 106–108
word order and, 59–60
Nowak, Martin, 207, 209, 210, 232
null subject parameter
children and, 194–195
continuous variation view and, 83
defined, 43–44
explained, 57–58
of Italian, 45
linguistic diversity and, 86
linguistic problems and, 49
ranking of, 167–169
Nupe language, 78–80

object. *See also* direct object
in compound languages, 146–147
extended polysynthesis parameter
and, 151
Mohawk language and, 99–104
verb-object constraint and, 93–97
object agreement
in Chichewa, 144–146
in compound languages, 146–147
effect of, 153
subject agreement and, 155
object noun phrase, 89–90, 144–145
object polysynthesis parameter,
149–150
object pronoun, 107
object-subject-verb languages, 128,
138–139, 174
object-verb-subject languages,
102–103, 138–139, 165–167
Oceanic languages, 129
Old English, 217, 220–222
Oneidas, 86
optimality theory, 78–79, 149,
151–152
optional polysynthesis parameter, 164
oral poetry, 234

Paiute language, 25–26
Paleosiberian languages, 115

parameters
adjective neutralization parameter,
176–180
agreement principle, 154–155
anthropological view and, 200–203
as atoms of "I-language", 57
as atoms of language, 52
children's language acquisition and,
192–195
defined, 58
effects of, 156
ergative case parameter, 180–182
evolutionary biology, linguistic
diversity and, 206–216
explained, 43–46
extended polysynthesis parameter,
150–153
formalists and functionalists and,
80–84
head directionality parameter,
68–75
hierarchy of, 161–171, 170–173,
183
language changes and, 217–222
levels of effects of, 125–126
lexical differences and, 79–80,
188–191
linguistic atoms and, 123–124
linguistics and, 49–50
mystery of, 227–230
need for, 19
Noam Chomsky and, 22–23,
35–36
null subject parameter, 43–44
object polysynthesis parameter,
149–150
optimality theory and, 77–79
as periodic table of linguistics, 158
phrases and, 65–68
polysynthesis parameter, 111–114
postmodernism and, 203–205
predictions about, 173–175
question movement parameter,
184–187
ranking principle of, 163
reflexive antecedent parameter, 187